Library of Congress Cataloging-in-Publication Data

Catalog record is available from the Library of Congress

LEWIS PUBLISHERS, INC.
121 South Main Street, Chelsea, MI 48118

PRINTED IN THE UNITED STATES OF AMERICA
1 2 3 4 5 6 7 8 9 0

BACK INJURY PREVENTION HANDBOOK

David W. Apts

LEWIS PUBLISHERS
Boca Raton   Ann Arbor   London

FOREWORD

"To love what you do and feel that it matters—
how could anything be more fun?"
Katherine Graham

Enough books and articles have been written about back pain since 1900 to fill an independent library. Volumes of scientific articles and books written by physical therapists contribute to the body of knowledge in the health care industry.

After spending 20 years in the field of physical therapy, and focusing on low back pain for over 15 years, I have arrived at many conclusions. Most importantly, back pain is no respecter of persons. World literature reveals back pain is man's most important nonlife-threatening disease. That statement comes from the opening of a book entitled *Idiopathic Low Back Pain*. This symposium of world leaders in back pain treatment, research, and prevention concluded in 1980 that most low back pain (85–90%) cannot be anatomically diagnosed. Furthermore, in *Spine* (Volume 7, No. 2, pp. 141–149, 1982), a synopsis of the symposium further states the "spine of man is not an evolutionary fault leading to backache."

Back pain is universal and its exact cause remains unproven. Reports show surgery for low back pain required for 9-year-old Japanese children and 80-year-old North Americans. Traditional medical approaches are being discarded and newer methods seemingly crop up daily. The old bed rest, pain pill, muscle relaxers, X-rays, hot pack, cold pack, ultrasound, diathermy, and even surgery have undergone close scrutiny and have shown there is no one right treatment or approach.

Having searched volumes of literature, I have come to the unmovable belief that back pain can and should be prevented at all cost. I have not attempted to write a scientific book laced with fancy terms and graphs. Instead, I present a methodology used over the past 10 years along with information about low back pain from world literature. As you read this text, I hope you will find it to be provocative and stimulating.

This book is a call to action. Its purpose is to help equip trainers, safety specialists and, therefore, industry contain cost and human suffering due to back pain. Although back pain is not life-threatening, it can change your life forever. It can rob you of life's many joys such as picking up your children or grandchildren to hug them, independently putting on your shirt and pants or being able to provide for your family. No, it is not life-threatening, it is life-changing. It sentences many to a life trapped in a painful body.

Please do not fight me mentally as you read this book. Relax and clear your mind of all previously conceived ideas and absorb what is presented. You may find it useful to reread certain chapters to obtain the intended message. With the strategies discussed in this book, you will be empowered to start getting back pain under control immediately.

DEDICATION

This effort is dedicated to the laborers of industry — the backbone of every nation, and to the thousands of employees I've trained in railroads, steel mills, coal mines, etc., for making my life worthwhile. You all have taught me that back pain is preventable and you helped make my job fun.

Thanks to Robin Deere, Laura Bustetter, and Paula Walsh for their endless hours of pouring over this book with me and for putting up with me. Also, thanks to the thousands of patients who have made this program a necessity by virtue of their suffering, to my staff at the American Back School and Physical Therapy Centers of America for putting up with me, my faculty for mentoring me, and my family for their love.

THE AUTHOR

David Apts became a Physical Therapy Assistant in 1973, and a Physical Therapist in 1977. His first job was as a Staff Therapist in a 350-bed hospital, then as Chief Physical Therapist in a 160-bed hospital. In 1985, he opened his own clinical practice in Ashland, Kentucky, which has grown into a 5,000 square foot, ultra-modern clinic featuring isokinetic computerized testing equipment, auto-traction, and all modern modalities. Today, this original clinic is one of a growing network of Physical Therapy Centers of America (all in direct access states) under David's supervision.

David is the author of numerous books, including *Back Facts*, which has sold nearly 2 million copies. His books, manuals and audio-visual programs have been translated into three languages and are used and endorsed by some of the largest companies and government agencies in the country, including:

Shell Oil
Peabody Coal Company
The U.S. Navy
The U.S. Air Force
Mississippi Power Company
The Chessie System Railroad
Norfolk-Southern

David, as President of Corporate Back Safety, a division of the American Back School, and his staff have been responsible for the back safety training of an incredible 2 million plus employees in industry, and he has been recognized in leading business publications including *Forbes* and *The Wall Street Journal* for his leadership in this field. He has presented more speeches and seminars than any other Physical Therapist in history. In addition to the seminars conducted for his corporate clients, David has been privileged to speak to the National Safety Council three times,

The American Coal Mining Congress, The American Society of Safety Engineers and The American Steel Health and Safety Congress, as well as many other associations. He is one of only two Physical Therapists to be members of the National Speakers Association. A 30-minute television program aimed at educating the public about back safety and injury prevention featuring David has also aired nationally. He is unquestionably the Physical Therapy Profession's #1 Public Relations Ambassador.

He has also conducted well over 120 seminars and workshops for medical practitioners, and his American Back School's 4-day Back Seminar has been the only program integrating the diagnosis and treatment of low back pain via categorizing signs and symptoms, musculoskeletal screening, patient education, and industrial consultation. He has helped thousands of Physical Therapists establish themselves as consultants to industry.

The American Back School, founded by David Apts in 1982, provides industry with books, manuals, audio-visual programs, video tapes, "Train The Trainer" Schools, preferred provider referrals and many other services. Complete information about these services is available free of charge by calling toll-free, 1-800-637-BACK.

TABLE OF CONTENTS

Dedication
Foreword
About the Author
Table of Contents

CHAPTER ONE

The price of greatness is responsibility.
Winston Churchill

A 56 BILLION DOLLAR PROBLEM

Many articles and books tell us that we in America have a 50+ billion dollar a year problem. Unfortunately, little has been done to curb this spiraling expense.[32] Presently, on any given work day there are over 10,000,000 people seeking treatment or are off work due to back pain. I submit the estimate of 50 billion dollars a year is too low.

The enormous loss of productivity at these rates is staggering. The National Center for Health Statistics shows the number of people disabled from back pain has increased by 168% in the U.S. from 1970 to 1986. That is 14 times faster than the population growth.[32] It has been reported in lectures and seminars some four million Americans are disabled due to back pain.

In other words, back pain, injury, or trouble is out of control. With all that has been written and studied about back pain, we still are at a loss as to its cause in many cases. In his presidential address to the North American Spine Society, Dr. Scott Haldeman reiterates that "we do not know the cause of back pain" and calls all of us to action.[32] He states that to get control of both disabling back pain and its costs will require a major

1

effort on the part of industry, government and the health care professions. I applaud his remarks and repeat that we will get a handle on back pain only with a team effort of all the players outlined in this book.

Miners (coal, zinc, gold, salt, etc.) in the U.S. are mandated by law to have 8 to 16 hours of annual retraining each year. It is mandated! Violations of the law result in financial penalties. Guess what. Miners get trained and educated on safe work practices. Perhaps industry at large needs to mandate this policy prior to the government requiring it by law.

BACK PAIN IS UNIVERSAL

Lest you think back pain is an American phenomenon, let me assure you it is not. As mentioned in a later chapter, there are those authorities who, years ago traveled throughout Thailand, Afghanistan, Iran, Iraq, Turkey, etc., and concluded that those nations do not suffer from backache. Nothing could be further from the truth.

In 1983, I was contacted by Mr. Joe Cullen, P.T., of Dhahran, Saudi Arabia. Joe was the chief therapist at the health center in Dhahran. He was looking for someone who could help train his therapists, treat patients, and implement industrial prevention programs. He wanted that person to come to Saudi and train his staff.

To make a long story short, I saw many foreigners as well as Saudi's who suffered from back trouble. Some of the problems I saw in Dhahran, I've never seen in the states. When I asked the medical staff if they saw a lot of Saudi's with back pain, they said that they were overwhelmed with them.

Even though Aramco brought me to the Kingdom of Saudi Arabia, I still had difficulty understanding why. You see, I believed the premise that people in that region of the world did not have back trouble. Not only did Aramco have problems with too many people showing up for back pain treatment, they also had it on the job — some 80,000 employees plus their dependents. So this company was responsible for the care of perhaps 250,000 people. They were being pro-active in trying to get a handle on the enormous cost of back injury.

I must confess, in 1984, my thoughts were still not as developed as they are today. We did not have enough of a track record or data at that time to develop a five-year back attack plan. Nor did I have the chance to meet with the proper officials of that company in order to initiate a plan of attack. However, working with the medical practitioners and patients was a truly rewarding experience. Saudis are warm and friendly people who graciously accepted me and our teachings on back injury prevention.

ITALY HAS PROBLEMS, TOO

In the fall of 1988, Dr.'s Luciano Merlini and Claudia Granada of the Rizzoli Orthopedic Institute in Bologna, Italy, spent two days in our Ashland Physical Therapy Center.

Through that meeting and seeing patients, together they sponsored me to go to Italy in October of 1989. While there, I not only had the privilege of lecturing hundreds of doctors and physical therapists, but also of seeing patients. What wonderful people and a beautiful country! Again, I found that the patients in Italy were not unlike the patients I saw in Saudi Arabia or here in the U.S. The point is that back pain is universal. I do not mean to imply that all nations and all people suffer from back pain, as I'm sure there is perhaps a group of people somewhere on earth that don't have back pain. The problem is, I don't know where to find them.

World literature abounds with back pain articles and books from Europe, China, Japan, North and South America, Australia, and New Zealand. The costs for all of humanity is anyone's guess. I can tell you that with increases in our work population, there is going to be more back trouble.

By planning today, we may start turning the tide on this epidemic. But it will require planning, and that's what this book is all about — empowering you to utilize a specific plan for you and your company and to help you get started now.

CHAPTER TWO

Great works are performed not by strength, but by perseverance.
Samuel Johnson

FAILURES OF TRADITIONAL
APPROACHES TO BACK INJURY

Historically, traditional medicine has dealt with injuries only after the fact. An individual had to be sick or injured before he would seek medical assistance; the medical establishment never gave any thought to anyone but a sick or injured individual. By placing all the emphasis on rehabilitation and no emphasis on prevention, the medical establishment and industrial leadership have made it difficult to contain costs.

Only in about the last 15 years have we seen pioneers in various medical fields and disciplines looking at the role of prevention. It has recently become somewhat popular, even trendy, to be involved in detecting people at risk for heart attacks, genetic disorder transmission, and other ailments; measuring their risk; and advising them on preventative and corrective steps they can take. The buzzword "wellness" has hit the marketplace with impact ranging from grocery store shelves to the hospital. Unfortunately, this trend has still not had major impact on the containment of health care costs in industry, or for that matter, in general. I recently read in the newspaper that America spent 500 billion dollars on health care in 1990 and by 1993 it will be over a trillion dollars! To a large degree, industry and the medical practitioners who serve it are still chasing problems rather than preventing them. Why?

Here's a case in point: quite recently, I introduced myself to a new doctor in my town and gave him my card. On my card is a line that says "Prevention Of Back Injuries." It completely baffled him! In fact, he laughed at the idea! The entire concept of preventing injury was beyond his scope of education and life experiences. That does not make him a bad person, not even a bad doctor. As a matter of fact, he has a reputation as a tremendous diagnostician. Unfortunately, he has a "mental block" about something of extraordinary importance to all of American industry and its work force. It is going to be up to industry to find open-minded, progressive health care practitioners to serve them, and to educate those practitioners about industrial life.

According to leading back specialist, author, lecturer and researcher, Dr. Malcolm I.V. Jayson of Bristol, England, "low back pain is among the the most common of human disabilities. Indeed, at some time in our lives most of us suffer from at least minor spinal symptoms. Nevertheless, our understanding of the problem is extremely limited. In the past, the subject was neglected by the medical profession, allowing scope for nonorthodox practitioners. The lack of interest was partly due to the very considerable difficulties, if not impossibilities, of achieving a precise diagnosis in many instances as well as to the feeling that often there is little constructive help to offer. It is fortunate that in the past few years enthusiasts have appreciated the importance of the problem."[43]

It is unfortunate that one quotation from the book is often misused. In Chapter 5, written by Dr. Allan St. J. Dixon, he refers to Fry's article on back pain and soft tissue rheumatism, which states, of patients not consulting their practitioner for low back complaints, 44% were better within one week, 56% better in one month, and 92% better in two months. Although that is a wonderful observation on the natural progression of back pain, we still do not know what "better" means.

It has also been stated that of this group of back pain sufferers, 80% will have recurrent symptoms. Obviously, someone is not getting better. Or should I say "trained" on how to prevent future episodes of low back pain.

Any health care practitioner who believes he knows what a person does on the job is deceiving himself. He cannot understand unless he personally visits the job site and studies the job and the work environment. Take any particular job in your plant. Can health care practitioners treating your employees answer these questions about that job?

1. How many repetitive movements are involved per minute, per hour, per work shift?
2. How much movement of weight is involved, to be lifted, to be carried, to be pushed, to be pulled?
3. How much weight is actually being lifted?
4. Are there injuries occurring at that job? How many injuries and how often do they occur?
5. When was the job site last analyzed for ergonomic appropriateness?

I began developing my methods for back injury prevention with railroads, steel mills, and coal mines. I have been most fortunate in that I went into about 25 mines, worked with the miners, and, as a result, understand their industry and its inherent problems. Fortunately, many in physical therapy have done the same thing and, as a result, have achieved a position in industry as respected, knowledgeable musculoskeletal injury prevention specialists. By taking knowledge out of the clinic and into the mine, factory, or work site, we can really make a difference. Did you know, for example, that most ergonomic job site changes can be done for under $500? A simple job site change, such as altering the height of a work bench, can prevent back, neck, knee, hip, or shoulder injury.

CHAPTER THREE

Having once decided to achieve a certain task, achieve it at all costs of tedium and distaste. The gain in self-confidence of having accomplished a tiresome labor is immense.

Arnold Bennett

PHYSICAL THERAPY: AN UNTRADITIONAL ANSWER THAT MAKES SENSE

Some industries may resist the influence of the physical therapist, sticking to the belief physical therapists must be supervised by physicians. Twenty-five states now provide direct patient access to physical therapists, and in twenty-one other states, physical therapists can evaluate without physician referral.[63] If you compare education given medical doctors and other health care practitioners to that of physical therapists, you'll see for yourself. It is the qualified, **experienced** physical therapist who is best suited to help you and your employees prevent back injuries and get related costs really under control.[6, 63] Let's take a look at why.

A PROFESSION IS BORN

Most people have no idea what and who a physical therapist is. Our profession came into existence during WWI, grew during the polio epidemic of the 1940s and 1950s, and has literally tripled in size in the past 15 years.

The American Physical Therapy Association (APTA) has adopted a philosophical statement that identifies movement dysfunction as the physical therapy content area of expertise.[63] There are some 17 subspecialties in the APTA. I am a member of the orthopedic, sports, and private practice specialities.

The physical therapist has been called on in an ever demanding role. An article by Shirley Sahrmann, Ph.D., P.T., on PT diagnosis reveals physical therapists' primary responsibility has been to understand physiology, pathophysiology, anatomy, and the components of kinesiology and kinesiopathology (the study of movement disorders) because that information is the basis of their practice.[63] Additionally, other professionals have little academic preparation in those areas.

Dr. Sahrmann contrasts the physical therapist's field of study to medical doctors, whose practice has moved toward a chemical basis. Their knowledge of molecular and submolecular structures is fundamental because of the pathophysiological basis of most diseases; gross anatomy has been de-emphasized. Further, Sahrmann states that differences in academic directions are part of the reason why physical therapists must become diagnosticians. The diagnosis is of movement dysfunctions, signs and symptoms, or clusters thereof. Physical therapists do not make medical diagnoses; rather, we must establish a movement diagnosis in order to establish a plan of treatment with expected outcomes.

PT WORK WITH MD

As Steven Rose, Ph.D., P.T., pointed out in his editorial "Musing on Diagnosis", the diagnosis made by a physical therapist is complimentary to and not in conflict with that made by a physician.[6,61] In a later editorial, Rose states he does not think making a diagnosis is essential for direct access practice. He states that we are not required to make a diagnosis, but that we must be able to recognize clinical signs and symptoms that mandate a referral to another practitioner, require consultation before instituting treatment, or clear the patient to begin physical therapy. He gives examples of how one makes a clinical decision by identifying clusters of signs and symptoms of problems outside the scope of physical therapy practice and the appropriate physician to which to refer patients, i.e., orthopedist, neurologist, and rheumatologist.[62]

For years before, Kentucky acquired direct access (which means, by law the public can see us without an M.D. referral just like a masseuse, YMCA instructor, occupational therapist, holistic homeopath, neuropath, etc., which can all see people off the street yet do not have as rigorous an education or licensure as physical therapists do), I received referrals from physicians, (as did my colleagues across the USA), with this type of a diagnosis:

1. shoulder pain
2. hip pain
3. low back pain
4. see & evaluate
5. stroke or CVA
6. burns
7. CP (Cerebal Palsy)
8. post-MI

Do you get the picture? I am not putting physicians down. That would be most foolish. I am attempting to show that physicians have used us for years to assist them as team members in rectifying their patients' problems. For example, as a physical therapist, I can make over 20 different musculoskeletal diagnoses for shoulder pain. Each will drive a different set of treatment strategies in the attempt to resolve the given problem.

PT'S WELL PUBLISHED

In their classic, timeless book, *Posture and Pain*, Henry O. and Florence Kendall have made a tremendous contribution to the subject of musculoskeletal dysfunctions causing postural problems and painful conditions. According to Robert Johnson, M.D. (Adjunct Professor, Orthopedic surgery, The John's Hopkins Medical School), "the Kendalls eliminate old false conceptions and add new concepts based on soundness of analysis and evaluation and in the clarity of their illustrations in their text." He stated they demonstrated perfect teamwork, rare technical skill, and great tenacity of purpose.[39] Since the publication of the book, there have been three editions to *Muscle Testing and Function*. These books show health care practitioners exactly how to assess and diagnosis certain movement dysfunctions.[38]

PREVENTION IS A MUST

Since 80% to 90%[54] of all back pain defies a definitive diagnosis, it has been to the advantage of the physical therapist to evaluate movement problems, daily activities, work conditions, etc., in understanding how they affect patients. Through repeated improper movement patterns, cumulative stress takes place and pain may occur. Conversely, if we correct abnormal movement patterns, weak muscles, tight muscles, and teach individuals how to care for themselves, we see back pain, neck pain, headaches, jaw aches, and many other bizarre symptoms disappear.

Many other references occur in literature on the contribution physical therapists have made in the treatment of low back syndromes. Donaldson, Silva, and Murphy have just published a beautiful article, "The Centralization Phenomenon and its Usefulness in Evaluating and Treating Referred Back Pain".[18]

ATTITUDE AND SELF CARE

Even Klein and Sobel, in their quest for a cure to their back problems, recognize the contribution of physical therapists.[41] Hailed as the ultimate second opinion for back pain sufferers, this delightful book tells who in the medical field is most likely to help you. The beauty of the book is that the authors point out that your attitude and your practitioners' attitude are directly related to outcome. You must care for yourself.

You see, that is exactly what PTs have been teaching for years. Only in the past 10 years, with the popularity of back school and wellness programs, have PTs gotten out of the clinic and dealt with the public and industry.

As in any profession, not all PTs are alike. I have taken over 60 continuing education courses since 1977 on back pain treatment and prevention. And I have given, sponsored, or directed over 120 seminars to P.T.s, M.D.s and D.C.s since 1982 through The Back Seminars, various State Chapters of American Physical Therapy Association, and other sponsorships while other PTs treat people who suffer from strokes, neurological disorders or burns.

Even with all of the focus on back pain, there still exist patients that I cannot help. I have shown you that physical therapy is a dynamic, scientific, helping profession. Many of us are well qualified to help you in your cost containment plans for musculoskeletal injuries.

TEAM APPROACH

My ultimate preference is for a true team effort between physical therapists, other health care practitioners and industrial management in a truly comprehensive attack on the problem of back injuries, but it will only happen in your company if you make it happen.

OBJECTIVE DATA

Along with the team approach, another important idea is the use of objective standards and procedures for measuring workers' fitness. Many workers, when asked by the doctor if they can go back to work, say "yes or no". They don't know whether they can or cannot until they try! Many people go back to work early or without sufficient preparation, so re-injury occurrences are unreasonably high.

Traditional models have failed to help industry or its workers get back problems under control. There is absolutely no reason to assume the traditional approaches will suddenly work better in the future. A different approach is called for.[56,72,76]

According to Wiesel, et al., industrial low back pain is manageable.[76] From their two-year study they conclude that:

1. Good medical approach leads to cost savings.
2. Use of standardized medical approach and terminology is necessary for consistent care.
3. A good medical record keeping system is essential to the performance of useful medical analyses for identifying scientific problems.
4. Unbiased medical surveillance leads to changes in behavior for both the patient and treating physician.
5. Most importantly, the outcome for most low back pain patients in industry is not as grim as previously perceived, if their medical management is approached in an organized manner.

Also in the study, treating doctors realized they were being watched by the authors of this study. The treating doctors also came into agreement with the authors that back surgery did not return patients to full duty and that there were only limited indications for its application to low back problems.

How refreshing it is for me to see such notable authorities as Dr. Wiesel, Feffer, and Rothman make such a contribution to the containment of industrial back injuries from the medical side. We are now seeing some of the most brilliant minds in medicine, research, science, health care, ergonomics, industry, government, and insurance companies working toward a greater understanding of low back pain, its cause, treatment, and prevention. It is truly an exciting time!

FAILURES OF TRADITIONAL INDUSTRY'S HANDLING OF BACK INJURIES

In the last ten years, as I have traveled throughout this country and around the world, consulting with industries' leaders and training their work forces, I have observed one simple, glaring fact: in most industries, very little thought, time, effort, and dollars gets invested in safety. I watch and listen as leaders scream "bloody murder" because they spend millions of dollars annually for this thing we call "back pain" — but I still see them doing nothing practical to change the situation for the coming years. It is not going to improve by itself.

CHAPTER 4

You cannot do anything by doing nothing.

Unknown

BACK PAIN IS NOT A COST OF DOING BUSINESS

A company with 3,000 or more employees in manual labor-intensive jobs will typically pay anywhere from one to five million dollars a year for costs related to musculoskeletal injuries, including back, neck, and shoulder problems; carpal tunnel syndrome; and tendonitis of the elbow. Five million dollars! Most people grudgingly accept this as a "cost of doing business" and only try to limit it by quota and dictum. If you accept this as a cost of doing business, you may lose.

Instead of accepting the idea of inevitability, accept the idea of preventability! Make strategic investments now in planning, education, job site changes, and other productive measures to bring injury cost down next year, the year after that, and the year after that. Over a five year period, investing considerably less than the cost of injuries, any industry should be able to virtually eliminate the cost of back accidents and injuries from its budget. That's right — I'm saying if you're spending five million now, you can spend half that much, maybe even zero, five years from now. But you are going to have to virtually abandon your traditional approaches.

In the study of The Chelsea Back Program, the Biltrite Company changed its attitude towards low back pain and back injuries. They initiate proper care, changed work sites found to be causing back pain and worked as a team to contain back injuries. Over a five-year period their medical and compensation costs went from $250,000 in 1977 to $1,200 in 1981.[25]

You can do the same if you work at it. We now have a tremendous number of industries that have realized even greater savings, both reported and not reported. More on that in a later chapter.

A tremendous number of books have been written about industrial low back trouble. Three of my favorites are

1. *Occupational Low Back Pain,*[2] edited by Malcolm H. Pope, Ph.D., John W. Frymoyer, M.D., and Gunnar B.J. Andersson, M.D.
2. *Industrial Low Back Pain,*[77] by Sam W. Wiesel, M.D., Henry L. Feffer, M.D., and Richard H. Rothman, M.D., Ph.D.
3. *Work Injury Management and Prevention,*[79] by Susan J. Isenhagen, P.T.

Each of these books makes a significant contribution to the principles of cost containment. World authorities on the subject have contributed chapters to the books. Various models are shared as to the treatment, prevention, and screening of low back pain. Prevention, education, worker selection, work place evaluation, work place design, and fitness are the elements that make this a total control concept.

I believe the cornerstone of a prevention program is appropriate training and education. Once employees understand the goals of the prevention program, I have found ergonomics and other programs are well-received. Our best ideas for job site changes, after all, come from workers who work that very job we're studying.

CALL FOR IMMEDIATE TREATMENT

However, if a worker is injured, Dr. Gunnar Andersson believes "a program of (1) diagnosis, (2) care, (3) rehabilitation, should be put into immediate effect." This is unrealistic perhaps, but optimistic certainly.[2] I believe this will be the model of the 1990s.

Consider traditional reporting of injuries as an example. The back is a large area. To report that a back injury occurred is almost meaningless. Was it a sprain or a strain?, of a muscle?, ligament?, joint?, disc?, right or left side? Did it occur immediately upon lifting, pushing, pulling, slipping, or falling? Did it occur in normal repetitive movement of a job, within an hour of a particular lifting, pushing, pulling or falling incident, or within a day? At what time of day did it occur, and on and on and on. We need comprehensive data to successfully play detective, to gradually identify trends, and to spot needed job site changes (see appendix for data collection form). Traditional industrial approaches have failed. Ignoring the problem fails. Blaming the problem on wimpy workers or goldbrickers fails. Sending the worker home to bed rest, hot packs, and pills solves nothing. We have to change.

THE NEW WINNERS' MENTALITY

Most athletes and coaches agree that there is a psychology to winning, a winning mental attitude, and it begins with the premise, or belief, that winning is possible.

Nothing has impressed me more in my experiences with industry than the overall lack of a winning attitude toward the cost of back injuries. Most industries and their leaders are coming from the premise and belief that winning is impossible. This mentality is costing American industry well over 50 billion dollars a year and, if unchanged, may ultimately cost our country a lot more, including the ability to compete in the world market at all![77] Estimates for whiplash injuries has been placed at over 5 billion dollars alone.[78]

I recently saw a top American auto industry executive state that about $700 per American-made auto goes to employee health costs. What is it going to take for us to wake up? That, purely and simply, is an outrageous cost factor, and a lion's share of it has to do with the attitude that has permeated our industries.

Our railroad industry has been nearly destroyed by back injuries and related litigation. There are law firms completely and totally living off lawsuits against the railroads! The whole thing is so out of control that one worker, with a simple knee menisectomy could get a million dollars in 1988 for such an injury. With a few notable and praiseworthy exceptions, leadership of the railroad industry has sat there, with a negative attitude about the whole thing, and let it happen. They have let lawyers steal their business out from under them, and jobs out from under their work force.[23]

Take a look at that industry, then go take a look in the mirror — are you making the same mistake? Is it only a matter of time before your injury costs so severely erode your ability to compete that bankruptcy or takeover becomes your only escape? The first step toward getting control of your injury problems and costs is to adopt the winner's attitude.

"BUT THERE'S NOTHING NEW UNDER THE SUN"

In the early 1980s, the idea of "the back school" began to spring up in different industries, in different countries, and by most anecdotal accounts, had a positive effect everywhere they took place. More recently, the Swedish Back School[80] has gained tremendous notoriety, thanks to the positive impact it has had on Volvo automobile workers. This is one of two known true controlled studies and it has proven two things:

1. Those educated about back injury prevention are likely to work longer without injury
2. If injured, those educated are more likely to return to work earlier, and then remain on the job longer.

Why? Because the workers learned how to take care of themselves, how to prepare for work, and were motivated to take personal initiative in preventing their own injuries.

MANY BACK SCHOOLS

This example has inspired a large number of "back schools" throughout the United States, Canada, and a number of other countries, including my own "American Back School" system", which has been used to train over 2-1/2 million people in industry as well as millions of patients in hospitals and clinics throughout the world. Most recently, the American Back School System has been condensed into a for-consumers video cassette.

A French physician, Delpech, is our first look at an original back pain center in the 1830s. From looking at pictures of his institute, not much has changed in what we do today, but more in how we do it.

One can say "there's nothing new under the sun" with back schools.[24,33,53] They are all similar in that they include basic information packaged with slides, videos, or other audio-visual materials. Most deal with what the human body is made of, how it works, what happens to the body and the back in bending, lifting, pushing, pulling and sitting, and provide exercises and other instructions intended to prevent injury and improve fitness.

While none of this is new, that's no excuse not to use it. In fact, there's danger in flitting from the very newest, hot "fad" to another. If you want a great example of the negative results of leaving the basics in the closet to gather dust while pursuing all sorts of fad-ideas, take a look at the problem-products coming out of today's public education system. Arguably the greatest coach in pro football, Vince Lombardi, is well-remembered for his relentless emphasis on the basics of the game.

The back education objectives must communicate clearly from the instructor to the employees. Here are some you should consider:

1. Back trouble can be prevented.
2. Only you can take care of your back.
3. Today you will learn what your back is made of.
4. How your back works.
5. You will learn by demonstrations built into the program.
6. You will learn the five reasons for locking your back when lifting, pushing, pulling, carrying, etc.
7. You will learn how to sit correctly and learn about the use of low back supports.
8. We will go over exercises and how to choose beds.

These objectives help the employee know what is expected of him. All of your training aids, videos, slide/tapes, workbooks, and home reference books should be of superior quality so as to ensure the best training possible.

Another controlled study of the effectiveness of back schools, conducted by Moffet, et al., concluded back schools are even effective for chronic back pain sufferers. What is encouraging to me with this study is that they teach one method of exercises and assume strong abdominal muscles protect the low back. Although I disagree with their exercise and lifting approach, they got very positive results. Most recent studies show abdominals to be of lesser importance than back muscles when lifting.[17,34,65]

So this goes back to my observations over 15 years ago that perhaps it is not so much what and how we teach a back school, but that we teach something to give participants hope and methods for coping. Moffett's approach encourages a positive outlook, discourages a passive, dependent attitude, while using a back school as a useful substitute for other forms of expensive and less effective treatment. I believe a back school that is solidly founded on basics is a giant step toward solving back injury problems in industry. Whether it's new or not shouldn't be an issue — it's effectiveness should be the only issue. It should address the methods of back injury prevention both on and off the job. It must be specific for your workers and not a canned program for best results.

I can tell you, based on my own ten years of experience consulting with industry, if you do nothing but deliver a decent back school to your people, you'll decrease your lost time back accidents by 20% to as much as 40% the very first year, as you will see in the following pages.

QUACKERY

According to a report by the Chairman of the Subcommittee on Health and Long Term Care, quackery is a 10 billion dollar a year scandal.[54] Let's not fall for every new fad that comes along. Let's not lust after a new batch of "bells and whistles" to spend our training budgets on. Instead, let's strive for thorough, comprehensive implementation of proven strategies. Choose that path and the first new thing you see under your sun may very well be a huge reduction in lost-time back accidents.

PROOF THAT BACK EDUCATION WORKS

I became frustrated with seeing one worker after another arriving at my clinic with conditions I believe could have been prevented, and I decided to go out into industry and institute back injury prevention programs. There was very little evidence that such an idea could work. In fact, most of my colleagues as well as most of the people I talked to in industry thought the idea was completely impractical. In 1981, I was able to convince the people at the Chessie System Railroads, a division of CSX Corporation, to try a preventative approach. I was seeing many of their employees as patients and had first-hand experience with their problems, so I was able to convince them that my foolhardy idea was worth a try.

TRAIN THE TRAINER IS BORN

Case History #1: Chessic System Railroad

From September, 1981 to April, 1982, I personally trained over 5,000 Chessie System employees. We used the same basic education agenda discussed in this book, although we varied it slightly to fit each different group of workers job demands. For example, bridge and building, track and maintenance-of-way employees received three-hour classes that covered the various tasks they must perform daily plus daily follow-up; clerical and engineering people got condensed two hour classes. The immediate results were sufficient to cause Chessie's management to proceed, the next year, with my training of seven of their trainers, so they could conduct programs for the remaining 30,000 employees. Thus, the "train the trainer" concept was born. These trainers did a wonderful job in educating their fellow employees.

Case History #2: Mississippi Power Company

The cleanest, best data I have seen comes from the Mississippi Power Company project. Mike Rogers, P.T., MNFF, and an eight-year faculty member with The American Back School, trained 1600 of their employees in two-hour classes, twenty to twenty-five people per class. Three years prior to this massive training effort, the company had 171 lost-time accidents due to back pain. Since the training, the company has gone an incredible seven consecutive years without a single lost-time accident due to back pain! For seven years, they have literally eradicated back injuries from their work place. Mississippi Power's shining example is extraordinarily important to the utility industry in specific, as well as all heavy and medium-heavy industries, and it absolutely proves back injuries at work can be stopped.

Case History #3: Peabody Coal Company

In 1985, I personally trained 27 people to be in-house trainers for the Peabody Coal Company, the largest coal company in the world. The 27 were trained 9 at a time, in two-day comprehensive classes. The "train the trainer" class format I use today is very similar to what these 27 trainees experienced.

The 27 trainees then carried the message to 12,000 workers in above and below ground jobs, in five divisions of the company. One year after this was completed, the company's Camps Division reported a savings through reduced back injuries of more than $500,000.00. I have not been privy to reports from the other four divisions, but even if their results were only half as good as was Camps, it adds up to about $1,500,000 returned to Peabody for an investment of about $20,000. I find this kind of wonderful return-on-investment to be typical!

Case History #4: Norfolk Southern

Norfolk Southern Corporation has sent over 120 people to be turned into trainers, and their most recent data indicates reductions in back injuries ranging from 50% to 90% in different areas of their company as a result of dedicated trainers' efforts.[1] They have consistently ranked #1 in safety in their industry, in large part due to the back attack program.

Norfolk Southern's success is especially notable because they are in the railroad business. In 1989, the railroad industry alone is estimated to have paid out over one billion dollars, mostly related to back and neck pain. The attorneys who litigate injury cases against the railroads have it down to a precise science. They also routinely entertain union officials and clients on their yachts and private jets. The special problems of the railroad industry (under F.E.L.A.) may yet destroy it.

You may ask where is the logic? The answer is simple, it will take an act of congress to change F.E.L.A. Who sits in congress … attorneys? Some, like Senator Howard Metzenbaum of Ohio, are partners in law firms that are making a killing. Do you think they will "bite the hand that feeds them"? Remember, you and you alone elect these officials.

What these attorneys fail to understand is that they are putting people out of work, increasing inflation and may terminate an industry. Who pays for the end result of all this litigation? You and I — we are the consumer! These lawsuits cause the railroad or any industry to increase their prices which are passed on to us, middle-class Americans. We are already carrying the burden of the tax system. We must not be complacent. We must vote for change and get these so-called leaders, corrupted by money, out of office. I'm all for the legal system, however, parts of it are out of control.

Yet even in such an environment, a company like Norfolk Southern is able to stand apart and take a road less traveled, for the benefit of its customers, the U.S. taxpayers, and its employees. They have been able to win the Harriman Award for the best safety record in their industry for three years in a row. Large legal settlements sound good on paper, but no amount of money can take a man's mind off his pain, his inability to work, and his loss of dignity and self-esteem. Preventing injury in the first place is best for everybody.

In just one of their departments, maintenance of way, lost-time back injuries have decreased from 98 to 47, from 1988 to 1990. Reportable accidents have gone from 142 in 1988 to 74 in 1990. Twenty-three accidents needed medical attention in 1988, compared to 13 in 1990. Total accidents and reportables have been cut in half.

Case History #5: AMAX Coal Company

In 1990, I received a letter from the Western Division of the giant Amax Coal Company as follows:

"In 1987, the Bel Air Mines of Amax Coal Companies received The Centennial Of Safety Award. This is jointly given by The American Mining Congress and The Mine And Safety Administration. Amax Coal Company also gives the "MAX" award. After training with The American Back School, the Bel Air Mines have received that award three consecutive times."

"We would also like to add that it is very seldom that a company will ever work one million hours without any injury occurring. After going through The American Back School's programs, we have worked one million consecutive hours without even one recorded back injury."[37]

All companies and organizations are not as open about sharing their internal statistics, and some, like U.S. government agencies, have set policies against doing so. However, when you consider the incredibly tough working conditions in coal, railroad, and utility industries, these successes should serve as convincing proof to lighter industries that they, too, can take control of their back injury problems.

CORE ELEMENTS OF SUCCESS

The core elements these companies have had in common and that have been most responsible for their successes have been acknowledging the scope and size of their problems, determining to solve them, developing a plan of attack, and dedicating reasonable amounts of time and money to carrying out their plans. Any company can copy that model! There exists many more articles in literature on the savings of injury and dollars due to back pain.[10,22,25,71,76,80] We must never discount the miraculous human body and mind that we have and its ability to heal itself when given proper instruction.

INCENTIVES

One other method that seems likely to help in cost containment for back pain is incentive programs.

According to Kendall,[40] team efforts within an industry are the key to managing this program and having fun. Each team that stays accident free for one month receives a certain number of points. These points are redeemable for merchandise from an award catalog. He states that it cut their incidence rate in half over a two-year period.

I have seen some of these programs work wonders and others fail. As in training, there have to be objective rules and common goals. Once started, it is best to keep a good program going. Perhaps a one-year time limit would be best. Then, changing to a different program will maximize compliance.

Many companies are now using dynamic low back supports as a safety gift. Whatever the gift, it should be meaningful, and would be nice, if it helped with their health or safety.

As you can see, there is a preponderance of proof that safety training and awards can and do make a difference on your bottom line of productivity and profitability. I believe a back school that is solidly founded on the basics is a giant step toward solving back injury problems in an industry. Whether it's new or not shouldn't be an issue — it's effectiveness should be the only issue.

I can tell you, based on ten years of experience consulting with industry, if you do nothing but deliver a decent back school to your people, you'll decrease your lost time back accidents by 20% to as much as 40% the very first year.

CHAPTER FIVE

Accept the challenges, so that you may feel the exhilaration of victory.
General George S. Patton

COMPANIES THAT CARE, EDUCATE

Putting people in a room for an hour, showing them a film, and sending them back out into the work place may technically qualify as delivery of a back school — but it's not true education.

Some years ago, I developed the model of delivering a back school to small groups of twenty to thirty employees, and actively involving them in the process. This naturally led to training trainers, so that a large company's employees could be divided among trainers at convenient times and places.

SCOPE OF BACK PAIN

People first need to be brought face-to-face with the true, mammoth scope of the problem, what it costs their company, industry, and the taxpayer, as well as its growing impact on the ability of American industry to successfully compete and preserve jobs. In other words, they have to be convinced of the severity of the problem and the significance of the cure.

Economic self-interest is interesting. Let's face it; if people were highly self-motivated to protect and preserve their health and fitness, we wouldn't see some factory areas filled with workers fifteen to thirty pounds overweight with their "bellies" hanging out over their belts! In order for a back school to work, it has to be aimed at the individual worker's self-interest. The health benefits have to be "sold" the best way possible. However, I've found economic issues get many peoples' attention.

BACK PAIN COSTS EVERYONE

If we take the estimated 56 billion dollar annual cost of back pain to American industry,[77] which passes it on to customers so that it ultimately reaches consumers via inflated prices, it figures to be an average of $1098 per household per year. In ten years, $10,980 will have been stolen right out of your wallet and right out of my wallet by this unnecessary, preventable cost! The following example is how the numbers were estimated:

USA Population/4.5 per family unit, 230,000,000/4.5 =
51,111,111 households or family units
Cost per year for Back Pain, $56,000,000/51,000,000 =
$1,098.04

Obviously, this is an assumptive example, however, it proves the point that back pain costs everyone over and above insurance premiums, worker compensation, and litigation.

I also think your back school should be presented to the employees as a benefit. That doesn't mean classes should be optional. They should be mandatory, but some effort should be invested in convincing participants that what is going on is of real, legitimate benefit to them, and can be used at home as well as on the job. They can share and need to share the information with their families and friends, as we have a "back pain epidemic" on our hands. Your employees will find they have learned quite a bit. You always learn more when you share with another.

BACK CLASSES CAN BE FUN

A real key is to make the classes genuinely interesting and dynamic. After all, nobody wants to come to a class about back pain. It doesn't sound like much fun. It's not sexy. It's not exciting. The employees' initial reaction to the whole idea will be fear of being bored to death. Unfortunately, in too many cases, their worst fears come true. I've seen back schools delivered in unintelligible, clinical terminology by instructors so expressionless you can't even see their lips move, lecturing in monotone, droning on and on. I've also seen classes where the instructor turns on a film and walks away. Would you tune in or tune out, if you were sitting there?

In my back school classes, everybody gets out of their chairs and involved in active demonstrations every ten to twelve minutes. Both visuals and demonstration objects are used. I avoid clinical language. The employees have the right to understand and, therefore layman's language should be used.

Another important concept that comes with training experience is bantering back and forth with the class participants. It is always done in fun, and some of the greatest humor I have ever seen surfaces during the training. If I had just one word to describe our success, it would be that training was FUN!

Training other trainers to do this has not been easy, but it's been very worthwhile. Those managers, safety officers, foremen, and others in industry who use these techniques know how to conduct interesting, dynamic, interactive classes that people actually enjoy. This, in turn, has better results.

TRAINER'S COURSE

Following is an outlined method for teaching trainers. My recommendation is for a two-day course, eight hours a day, in order to thoroughly train, explain, demonstrate, and empower those trainers to get results. And, yes, trainers do get results as we have already seen from previous testimonies.

28 KEYS

I give them 28 keys to success. For example:

- Be on time. It allows you to mentally imagine how the training will go and gives you time to mingle with the class participants and break the ice.
- Have mandatory eye contact with your audience at all times.
- Be physically prepared to show the class how to bend, lift, push pull, carry, sit, exercise, etc.
- Set the objectives of the meeting.
- Always be positive.
- Deal with that one bad apple and conduct hands-on time for dynamic interaction.

It is appropriate to cover proper public speaking techniques and how to tell stories, joke, and roll with the humor that emanates from the class.

Each trainer should be told several times that each class has to be better than the one before and that during those two to three hours of class their job is to ensure it is the very best class the employees have ever been to in their lives. Spaced repetition is powerful and should be used in training trainer as well as employees. An overview of this program can be found by David W. Apts, PT.[7]

WE ARE GOAL ORIENTED

We as human beings are wonderfully made. We are made to succeed and are goal-achieving people. Ask yourself right now how many teachers or professors made an impact on your life. Go ahead and think about that right now for five minutes and list their strong points.

What did you come up with? Check your list with mine. Good teachers:

1. knew their subject
2. loved to teach
3. knew how to push you to get more out of you

4. respected you for you
5. made their subject come alive
6. made their subject fun

You see, the right kind of humor helps people receive and remember information better. Some of our greatest statesmen had a terrific sense of humor. Although I do not remember Abraham Lincoln (my teenage children would say different), I have read a great deal about him and humor was a tremendous tool for him, as it was for Winston Churchill and John F. Kennedy. Look at the greatest comedians of our time and you will see their insight, wit, and sarcasm contribute to their longevity, such as George Burns, Bob Hope, Johnny Carson, Jack Benny, and Bill Cosby. Humor is one of the greatest tools we as trainers have and should use to the maximum.

CUSTOMIZATION IS A MUST

Another key is to carefully customize the back school to your work force and their activities. If you stick a wood box in front of your people for them to practice proper lifting, they will silently, immediately discount the whole process. In their minds they say: "They don't even understand what we do out there." The objects they work with in class need to be the same objects they work with on the job.

PRACTICE HELPS

When I worked with coal mines, for example, I made sure we had several five-foot timbers, five-gallon oil buckets, sixty-pound bags of rock dust, hunks of cable, and header boards. Then we practice the lessons, buddy lifting, Olympic model lifting, etc., with these "real" objects.

Here's the message behind all this: there's no shortcut to success with a back school. You can't just buy a box of "stuff," run some quick meetings, and say you've done it. You can't just bring any doctor, therapist, or lecturer off the street, slap him in a room, and let him lecture at your people. Some real thought, planning, concern, and time must be invested for real education to take place.

PRACTICAL HELP

The following are safe, simple exercises and rest positions that can be given during training sessions. Your trainers should be able to demonstrate each of these with correct cadence and technique. We are not trying to turn your trainers into practitioners. However, they must know proper procedures for exercising.

EXERCISES

REST POSITIONS

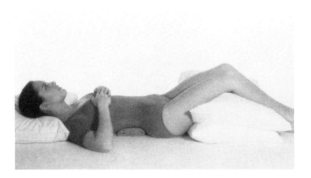

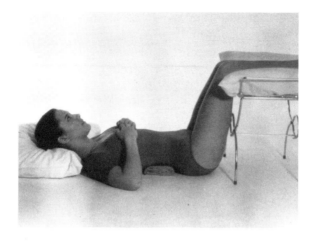

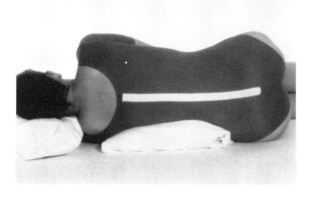

All rest positions should be used for 15 to 20 minutes only, then change posture or position. If you are stiff or hurt when changing positions at 15 minutes, you should change positions every 8 to 10 minutes. It is never wise to stay in one position too long. Some people tell you to sleep 6 to 8 hours flat on your back with pillows under your legs. However, this is not advisable, as it is healthy to turn frequently.

CHAPTER SIX

Luck is what happens when preparation meets opportunity.
Elmer Letterman

THE KISS METHOD

Your trainers can learn, in two days, everything they need to know to conduct dynamic, interactive, interesting classes. Topics cover a wide range, from our philosophy and purpose, to biomechanics of the back (how it works), choosing a bed, exercises commonly used for back pain, how you successfully conduct a class, and a checklist of everything they need.

Each seminar should run from 8 a.m. to 5 p.m. each day and has a class size limit of 30. It is intensive! With this progressive learning system put together into blocks of time, it is easy at the end of this course to assimilate the information and immediately put it to use upon return to their job.

EDUCATION — KEEP IT SIMPLE, SAFE

As with any training, it should be dynamic, interactive, and an intensive 2-day, 16-hour experience. Each trainer learns about the scope of back pain in the U.S. and abroad. They will also learn of the success of other industries that have worked practically to control back pain. In other words, each trainer learns that they are beginning a worthwhile adventure. The philosophy and purpose are explained so we can achieve a common goal.

The entire 16-hour course is instructed in layman's terms. I learned a long time ago that using medical jargon impressed only me. In order to communicate, everyday language is a must.

From the outset, your trainers are going to understand, we believe, that the human spine is intended to function in the upright posture with its three normal curves. These curves should be maintained while lifting, pushing, pulling, and carrying a heavy load.[8]

We regard the back as any other joint in the body because normal or functional flexibility and strength should be maintained, and because each individual should know how to use and protect his/her back.

We cover the fact that the initial massive attack is going to be successful, and we depend on the enthusiasm of and effectiveness of each trainer. From there, we cover food for thought. I have taken excerpts from Zig Ziglar's book, *See You At The Top*,[81] which have meant a great deal to me in gaining insight as to how we perceive each training class.

TREAT THEM RIGHT

To get things started on the right track, we as trainers must understand our responsibility. Zig says, "We treat people exactly as we see them." Now that is a mouthful.

I've included this saying, because I have, upon many occasions in the infancy of my consultation with industry, heard the following: "You think you can train my people for 2-3 hours? Who do you think you are? Why, when any of our upper management talks to them they are asleep in 15 minutes." No wonder, the way the meeting(s) were conducted, anyone would be asleep in 15 minutes. How about this one, "You can't train people who are dumber than coal buckets."

Well, if you believe they are, then they are. Personally I do not believe people are that dumb. Lazy, unmotivated, and rebellious at times, yes. Dumb? No! Look at the ingenious ways people have learned to get out of work. They rob, steal, cheat, and lie quite creatively, even though it's negatively directed.

I always tell trainers that their attitude for each class has to be the following:

1. This is the best class I will ever teach.
2. These people need to know what I will teach them.
3. An ounce of prevention is worth more, much more, than a pound of cure in the case of back pain.

If somehow we believe our class to be stupid, we can talk all we want but through nonverbal communication, we will tell them how we truly feel.

Each trainer must have enthusiasm, knowledge, and skill at what they are presenting and with the right attitude they will do a tremendous job for your company.

DESIGNED FOR SUCCESS

Another Zig saying I like is this: "Man is designed for accomplishment, engineered for success and endowed with the seeds of greatness."

Do you believe that? I hope so. Just look at history. When an enormous task is before us, we can reach into our inner depths and come up with amazing discoveries, put men on the moon, bring them home safely, win the World Series or Super Bowl, and put an end to a maniac like Hitler.

We presently have a task larger than any one of us, larger than any major corporation. In fact, it's even larger than America. It's called eradicating back pain or back trouble. I do believe we are equal to the task of solving this epidemic, or at least setting the stage for future generations.

Each trainer is put through the three-hour training program and tested on their newfound knowledge. The next several hours are directed to ward understanding each section of the program and how to deliver it in part and total.

The following are the parts that create a three-hour training program:

1. scope of the problem
2. how the back gets injured
3. five reasons for locking the back in
4. sitting — the silent killer of the back
5. lifting — Olympic style — research
6. activities — encountered daily at work
7. rest positions
8. exercises
9. choosing a bed
10. proper closing

To give each trainer more in-depth knowledge we also cover anatomy of the neck, torso, low back, biomechanics (how the spine moves), kinesiology (study of human motion), pathokinesiology, and complex back trouble that can occur.

Other topics such as surgery, chymopapain, MRI, CAT Scans, and conservative therapies, are briefly covered as well. Why? Dale Carnegie states that when preparing a 20-minute speech, you should amass enough material for a 2-hour speech. Because of your preparedness, you will be more likely to answer questions, and thus look like you know what you are talking about.

Also, we cover the pertinent elements of public speaking. The last part of day two is devoted to understanding the five-year plan and to having trainers give one of the ten segments that comprise the program to their peers.

This helps them overcome any jitters about using "new material" and gives the class a chance to offer constructive critiques.

COMPLETE KIT

Over the years I have learned that trainers, safety officers, and supervisors are busy people. It may be best to look for those programs that have turnkey trainer kits for their ease and convenience.

What you should look for:

1. slide/tape lecture — two to three hour program
2. manual — with complete instructions on how to use the program and incorporate the demonstrations and text for each slide

3. bibliography — search of world literature with author, reference, source, and critique
4. life-sized simulated human spine — a beautiful tool to show how the spine works
5. herniated disc — simulated ruptured disc
6. fifty-minute video — training highlights filmed during a train-the-trainer course in 1990 so that each trainer can plug it in and see how to conduct a class again and again
7. a dynamic low back support
8. back Books — correlated to the lecture program, pictures, tests, etc.
9. neck Books
10. spine and muscle posters — perfect for demonstrations, explanations and displaying on training room wall
11. information that can be made available to the employees family through literature and video presentations.

All the bases must be covered in order to effectively train your trainers. Giving them proper materials to present to employees and to use for their own review, equips the trainer with ammunition to get the job done. Between the variety of audio/visual materials, class demonstrations and participations, you will have covered all the different learning methods. Let's face it, some people learn best by reading, others by watching, some by hearing, and some by doing. Your training should encompass all areas.

CHAPTER SEVEN

The highest reward for a person's toil is not what they get for it but what they become by it.

John Ruskin

TRAINING THE TRAINERS

In 1983, Chessie System Railroad wanted me to train 30,000 employees in a short period of time so that we could closely monitor and measure the effectiveness of the back school. Because I devoutly believe in limited class size, I was going to have to conduct more than 1,000 classes, more than 3,000 hours. If I worked exclusively for Chessie eight hours a day for 365 straight days, I'd barely get it done. We decided, instead, to train trainers and replicate myself throughout their company.

Individuals were selected from each of their seven divisions and, in a week-long workshop, prepared to conduct a back school. Since that time, I've refined that process into a two-day course for trainers, and have utilized it for many Fortune 20, 50, and 100 companies and government agencies.

Many positive by-products occur with this approach, including some I never anticipated. By having qualified trainers in their own work force, companies can do a much better job with training thanks to total scheduling flexibility. If a class needs to be conducted for the night shift, no problem. If one shift or plant location is having continuing injury problems, the in-house trainer can go there and conduct refresher courses. These same trainers can get to the work sites and monitor compliance. These same trainers can present safety talks and lead ergonomic committees. While I originally went the train-the-trainer route out of necessity, I've since concluded it is the best approach for all but small companies.

Through turning people into trainers, I've found the first obstacle to be in their own minds. Many find it impossible to believe that the people they've been working with for years will suddenly accept them as authorities on back trouble. Actually, it is a very legitimate concern, and the only way I know to counter it is to match the in-house trainers with a back school provider that has strong credibility.

The trainers who go through my program graduate with diplomas that they can display as desired, and are able to go back into the work environment with the credibility and confidence of certified instructors for The American Back School. They can talk about the 2.5 million people already trained, and the successful case histories. Whatever source or sources you use for a back school and for the training of your trainers, you should find ways to transfer the credibility to your trainers.

Enthusiasm makes the difference. Trainers need to become enthusiastic about their work. Mechanical, robotic trainers simply don't get results. On the other hand, genuine enthusiasm is contagious. Trainers who take it seriously see how beneficial it can be, not just to their employer but also to their co-worker friends. They can take pride in their new role, having immense impact on your work force.

EDUCATORS ONLY

Your trainers must also understand and accept their limitations. The last thing you want is a trainer running around "playing doctor." They must know what questions they can and should answer, and how to answer them. They must also know what questions should be referred to an appropriate medical practitioner. All this should be covered during the seminar and in the manual. Developing a core group of knowledgeable, credible, confident, enthusiastic trainers within your own work force can be one of the very best human resources investments in the history of your corporation.[1,22,37]

All the methods I've used in industry over the past 11 years have been refined to a format of two to three hours. To this day, when I give a seminar to 25 to 30 industrial employees, I use this very format. I have personally trained over 150,000 people and know what works and what doesn't. Remember, NOBODY BUT NO-BODY wants to go to another back safety class. I know when to use humor, when to be serious, and where in the program we need dynamic demonstrations. It is my personal belief that the employees need to be up every 10 to 12 minutes participating in exercises, lifting, pushing, pulling, carrying, and material handling demonstrations. That way, nobody has time to sleep or mentally checkout while still physically present. I teach each trainer the little nuances and tricks I've picked up over the years.

It is structured from beginning to end so that it moves rapidly and is fun and interactive. Each trainer should train the same way. I want them to hear what I've heard from the past participants: "This is the best class I've ever attended."

Spaced repetition is a must both in training trainers and having them, in turn, train your employees. Many studies have established the fact that the human mind is like a sieve; not that we are dumb, but that we do not retain new information well. In fact, some authorities believe that 25% of a new message is gone in 24 hours and 98% is forgotten in two weeks.

Throughout this program, I carefully cover all the bases several times. Your trainers will do the same when they train their fellow employees.

TESTIMONIALS FROM PAST CLIENTS:

"… A lot of myths were shattered, a lot of new, beneficial information was learned! This information can be worth a million dollars a year to my company!" John O'Brien, Mine Inspector, U.S. Steel Mining Company

"… Best presentation on back injury prevention I've ever attended!" Donnie Coleman, Owner, Southern Safety, U.S. Steel Mining Company

"…Gave me a basis for formulating an effective back injury

prevention program, built around truthful and believable data (not hypothetical)." MSGT Rocky Cowart, U.S. Air National Guard

"... One of the better courses I have attended. It made me more aware of the problems the employees encounter in their day-to-day work." Bob Anglin, Georgia Pacific Corporation

"... This gives me and the company credibility in providing prevention rather than repairing the back after an accident." Gerald Mountain, Supervisor of Employee Relations, Consolidation Coal Company

"... This was a very professional presentation, but still down to earth. This will certainly benefit our company economically, but even more important is the humanistic viewpoint — caring for our fellow workers. I wish someone would have given me this information five years earlier, before I hurt my back!" Jim Ellenberger, AMAX

CHAPTER EIGHT

No problem can stand the assault of sustained thinking.

Voltaire

THE RIGHT LIFTING MODEL

Nothing incites more controversy in the field of back pain than the topic of lifting. Personally, I think it is foolishness. There is never one right way for everyone to do anything. To give you a little historical perspective, I first entered the health care field in 1970. I was taught to do a posterior pelvic tilt to lift, in other words, round the lower back out maximally. This was thought to be protective of the back. The posterior pelvic tilt has been taught in the medical field for over 50 years now.

LIFT LIKE A MONKEY OR LIFT LIKE A MAN

First of all, the human being did not evolve from apes,[92] and we are made with three curves in the spine, not two. Following the old medical model, you lift with an inward curve at the neck and a large outward curve from there to your pelvis. That position was thought to utilize the weak back ligament (posterior longitudinal) and hold the discs in place while the abdominal muscles supported the spine. That is the medical model for lifting. It is appropriate for some people, but they are a small minority.

TRY IT

You ought to try getting into that position (medical model) to lift an empty box or chair. You will find it nearly impossible to assume the position for any period of time and totally impossible to lift a weight of any magnitude.

I learned the method well and taught it daily for over three years before questioning its efficacy, and continued training patients that way for another seven years because doctors and bosses told me to do so.

Einstein has been quoted as saying if you learn something wrong one time, it will take eleven correct teachings to change your mind. I don't know if this is true; however, I do know that when you believe in a theory, even an erroneous theory and reinforce it daily, your mind will begin to accept the absurd as truth. If it were not for the failure of the medical model for lifting, I may have accepted it as truth to this day.

After teaching on that method for over a decade, I began to question everyone in my profession, but came up with no answers. All the doctors I knew were trained as I had been, so they could not help either.

ROBIN MCKENZIE

A wonderful physical therapist, Robin McKenzie of New Zealand, was touring the U.S. teaching his principles when I had the good fortune of meeting him in 1978. For me, he was the first to shed new light on the subject of lifting, and his clinical empirical observations and methodologies stand today.[47]

He stated his observation of Canadian Olympic lifters was that they lift with lumbar lordosis, or with the spine in its three normal curves. After taking his courses and employing his marvelous techniques, I found I could help patients with low back, buttock, and thigh pain who, for ten years, I had not helped.

OHIO STATE LIFTERS

This got the better of me and off I went to Ohio State University in search of their weight lifting team. Again, they told me that they lifted with the back "locked in". With the back "locked in", they could lift more weight and not hurt their backs.

As I was about to leave, another lifter came in. He was a graduate of OSU but still worked out with the team. He told me he used the principles of locking the back in at work. He worked at a ball bearing factory and some of the bearings weighed 50 to 100 lbs. When they were defective, they had to be lifted manually off the assembly line. He further stated that all the guys he worked with were older than him and laughed at his technique, but that they all have serious back trouble, and he said, "I don't and I won't because I know how to protect myself."

ENTER CURT CARDINE

Shortly after that, I was on a flight from Hartford, Connecticut to Pittsburgh, Pennsylvania and happened to be seated next to a rather large fellow. I greeted him with a "hi." He replied "hi" and buried his face back in his book. Judging by the size of his legs, I thought he was an NFL star and asked him if he played pro football. "No!" he said. Obviously, this guy didn't want to talk. So, I said, "Well, you're something because those are biggest legs I've ever seen." He then introduced himself as Curt Cardine, Olympic weight lifter. Curt is now a school principal in New Hampshire, has a master's degree in education, is still competing nationally in the 242-pound weight class, and is coaching over 20 teenaged Olympic style lifting hopefuls.

OLYMPIC LIFTERS LAUGH AT MEDICAL ADVICE

After I told him my story as a physical therapist who was frustrated with the medical model of lifting, he looked me in the eye and said that he and his fellow lifters (2,000 registered Olympic lifters in the U.S.) laugh at us in the medical field when we try to tell them how to lift. He told me that we didn't know what we were talking about.

About three months later, Curt joined me and Dr. Steven Joel Rose, Ph.D., P.T. as the core faculty of The Back Seminar. In two more years, we would have ten faculty members. I never heard health care professionals in our seminars tell Curt he didn't know what he was talking about. Yet they would not believe me when I'd tell them they should teach lifting with the back locked in.

TWENTY YEARS OF DRUG-FREE LIFTING

At age 39, Curt has been in international competition for 20 years, is drug free (never used steroids), and never had a day of back pain. He is as flexible as any world-class gymnast, swims and bikes for aerobic fitness, lifts three times each week, and has a 42-

inch vertical leap. Curt is a delightful educator and has helped make a significant impact on health care practitioners' attitudes about strength, flexibility, and, of course, lifting.

This has been a lengthy scenario because lifting styles are steeped in tradition rather than based on clinical observations, research, or studying the Olympic lifter. Can you believe that nobody from medicine or other health related fields has studied the Olympic lifter?

In 1984, at the Challenge of the Lumbar Spine in New Orleans, Dr. Serge Gracovetsky from Montreal, held fast to his theory of lifting entitled the Mechanism of the Lumbar Spine.[30] The theory is based on the idea that ligaments are what protect the spine during lifting because back muscles are too weak. Therefore, he stated the best way to lift is with a posterior pelvic tilt or back rounded out utilizing a hydraulic amplifier effect of ligaments, abdominal muscles, and the intraabdominal pressure.

Since that time, many researchers have found errors in the mathematics of Dr. Gracovetsky's formula. Dr. Ian Stokes[83] of the University of Vermont, and fellow Canadian Stuart M. McGill, Ph.D., of the University of Waterloo, Ontario, have found errors in the estimates of single moment arm length of the back and size of cadaver muscles in various lifting theories. They found with CAT scans of live human subjects, the moment arm length between the centroid of the back muscle mass measured 7.5 cm in back of the center of the disc. Most other studies thought the lower arm length to be 5.0 cm.[46] Imagine comparing cadavers to live dynamic human beings!

OLD MODEL PREDICTS TOO MUCH COMPRESSION

With this new detailed geometric modeling by McGill, we see predictions of compression on L4/L5 may be up to 35% less than the commonly used 5 cm moment arm model in older lifting theories and mathematical models. Also in the report, Dr. McGill states that rounding of the back as Gracovetsky and Farfan report was not observed in their lifters during strenuous lifts or in the studies of others.[46]

I have been told by some of my Canadian colleagues that I

would have to change the back-locked-in theory once I heard Gracovetsky and Farfan speak. However, I'm afraid that even Gracovetsky and Farfan are the ones who must change.

ABS GOES TO CANADA

After a two-day seminar in Montreal with Curt in 1984, I received a call from a man whose voice I did not know nor could I understand his language due to his accent. After about five minutes, I realized I was talking with Professor Gracovetsky. He said he had a copy of my book *Back Facts,* (copyright 1981), and that I was wrong in teaching people to lift with the back locked in. Furthermore, still yelling at me, he said he was going to put me out of business at the Challenge of the Lumbar Spine in New Orleans, November, 1984. I only wish I had recorded the conversation as he appeared to be incensed with me (and I am being most kind to the professor).

I marshalled my faculty and Curt Cardine called his friend Bud Charniga, Jr. to join us. Bud is a former Olympic lifter, now a coach, who has been to Russia four times and has translated the *Russian Weightlifting Yearbooks* [15] and *The Snatch and the Clean and Jerk* by Roman and Shakirzyanov.[58] All of the Russian literature show their top performers. When lifting monstrous loads, you do not see them rounding out their backs, as Gracovetsky would have us believe. Some people will see the truth, yet not believe. In the book, *Clean & Jerk,* there are hundreds of stop-action pictures showing the lifters through entire lifts with 11 to 17 snapshots per lifter. Even an untrained eye can clearly see that these men keep their low backs locked in.

We were never able to debate, let alone reason, with Professor Gracovetsky. However, the literature has upheld the Olympic model for lifting. How could our Olympic lifters and 360,000 Russian Olympic lifters be wrong? Can you now begin to get a feel for what I said earlier about erroneous hypotheses or theories? If reinforced, they can be tough to change. I am not saying it is bad to have theories or hypotheses — they are needed as we search for new understanding. However, erroneous assumptions believed as truth can be harmful.

In another delightful paper, McGill and Norman show previous low back modeling studies have greatly underestimated the role of back muscles in supporting the spine, and that the popular conception that reduction of disc compression takes place by intra-abdominal pressure is highly overestimated.[46] So, two strongly held theories must move aside. It is refreshing to see that recent research is closing in on answers for how the magnificent human body operates.

I would be remiss if I did not share with you at least three more studies on this issue of lifting.

BACK DESCRIBED AS AN ARCH

Recently, Aspden, et al., in the *Journal of Biomechanics,* have proposed a new mathematical model for understanding the spine. He calls for us to abandon the cantilever approach (which describes the spine as a crane) and embrace idea of the spine as a masonry arch. By describing the spine with three curves and using geometry, he clearly states the spine is strongest when held with its three normal curves.[8]

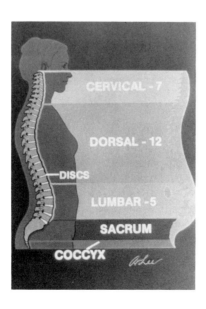

NEW RESEARCH IN LIFTING

A close friend, colleague and faculty member of The American Back School, Dennis L. Hart, Ph.D., P.T., wrote his dissertation on "The Effect of Lumbar Posture on Lifting."[34] He found that back muscles work most effectively with the back locked in and least effectively with the back rounded out.

Similarly, Rose, Delitto, and Apts report in their study that the back muscles work best with the back locked in, especially at the beginning of the lift. Our hypothesis was that back muscles should be working at greatest efficiency at the beginning of the lift. According to many studies the greatest stress on the lower back is during the first 200 milliseconds.[36] This seems rational as it is at the beginning of the lift when we are overcoming gravity. Abdominal muscles only work at less than half of their capability.[17]

This graph shows how much the back muscles work during two styles of lifting. Each style is a percentage of 100% of the back muscles maximum voluntary contraction. Our hypothesis is that the lower back needs to be under the control of muscle during lifting, especially the first half of the lift.

As you can see, the back muscles are working up to 70% of maximum with the back locked in, and 45% with the back rounded out at the beginning of the lift. Perhaps once the object is in motion, the greatest strength requirements are overcome and the body

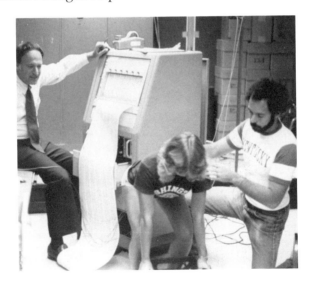

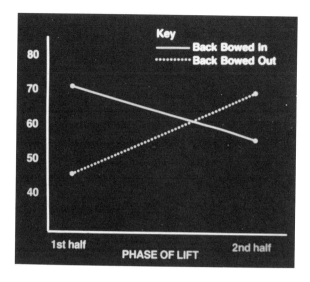

conserves its energy. So the muscle firing (EMG) decreases through-out the lift with the back locked in technique as opposed to continually fighting to stabilize the spine throughout the lift with back rounded technique. I know of no studies that correlate EMG activity with strength. All this study can tell you is that the back muscles work more with the locked in method.

Shiro compiled data similar to Hart and Delitto in her master's thesis at Columbia University. Her research was done totally independent of the American Back School and Hart, yet her results reveal that the back muscles work more effectively during lifting loads with the locked in technique.[65]

METHOD

Our test subject is hooked up to the EMG machine in the background. The EMG electrodes are placed on her back and abdominal muscles are record their activity during lifting. Male subjects lifted 25 and 50 lbs. while female subjects lifted 15 and 30 lbs.

Although there has never been any correlation between EMG activity and strength, we will agree on the EMG activity as a percentage of the back muscles maximum voluntary contraction.

As you can see from these graphs, there is more back muscle

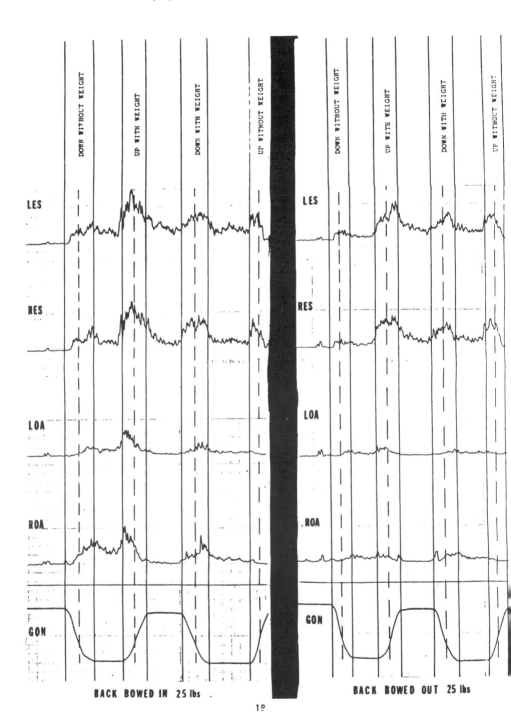

BACK BOWED IN 25 lbs

BACK BOWED OUT 25 lbs

activity when lifting in the beginning of the lift with two subjects who have their back locked in, both on the way up with the load as well as the way down. It is our belief that the spine has to be under active conscious control when under a load of lifting, pushing, or pulling by use of muscle. In fact, that is all you or I have to protect the spinal bones, ligaments, nerves, and discs. Ligaments cannot protect you, as they are not under your active conscious control.

Obviously, I could go on and on about lifting. It deserves our attention because one third to two thirds of all back pain can be attributed to heavy lifting. According to certain studies, Hult and others have found that heavy lifting is a risk factor for low back trouble.[17,32,36,44] Hult has found that up to 64% of workers who had back pain attribute it to lifting.

COAL MINERS AT RISK

In another investigation of coal miners, Rose and Apts found 92% of underground coal miners work daily with back pain.[59] We see that the unsupported spine in bent-over work postures is a risk for low back trouble. Lawrence, et al., found an increase of low back pain in miners in Great Britain, and attribute this to bent over (at waist) work posture.[42] One might wonder about the body's function when bent over.

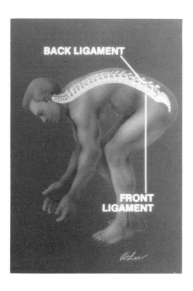

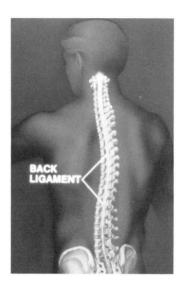

BACK MUSCLES SHUT OFF

Since 1952, studies have agreed that when one bends over at the waist and reaches full forward bending of flexion, the back muscles cease working.[21,29,30] Golding found standing toe touching (in normal people without back pain) caused back muscles to shut off. However, in back pain patients they do not shut off. That is a whole other topic about why we need proper post-injury training so that the body will recover fully from injury.

With the spine fully flexed, we are literally hanging by our ligaments. According to Rissanen, he found the interspinous ligaments to be over stretched or too weak to check forward bending in surgical patients for discectomy.[57] Panjabi, et al., found the greatest stress to be on the more superficial ligaments (supra- and interspinous ligaments). However, the stress works from superficial to deeper layers of ligament.[74] The backwall of the disc is considered by some to be the innermost ligament of the back.

So, the theory is that when one bends over at the waist often enough, lifts heavy weights, and twists frequently, the ligaments are subject to too much stress. If the tensile force is too high, the ligaments will start to break down and disc prolapse may occur.

DISC PROLAPSE

Rissanen states that, at surgery, a prolapsed disc is almost invariably associated with a ruptured or slack interspinous ligament (the ligament between the vertebrae near the surface of your skin. You can feel this between the bones of your own back). Is this how the body breaks down and leads mankind to disc rupture? I do not know for sure, but it paints a nice research and clinical picture. Remember, ligaments are not under your active control.

I have gone on at great lengths about lifting. In fact, one could write a rather lengthy book on the subject. My attempt has been to show you the disagreement (which is healthy) within medicine and share actual literature, old and new. I have believed in the Olympic model since it was shared with me in 1982. This model for lifting has not changed in over 60 years.

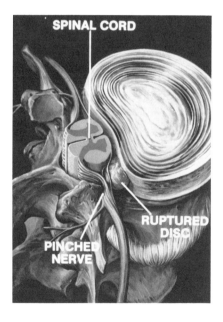

THE OLYMPIC MODEL CAN WORK FOR YOU

Please let me be bold enough to submit that your employees will never be lifting 300, 400, or 500 lbs. overhead. If Olympic weight lifters can train for 15 to 20 years and not suffer from back aches, perhaps lifting heavy weights is not a risk factor for low back trouble, after all. Perhaps our inappropriate movement patterns poorly recruit our musculature, and we, therefore, put ourselves at risk without even realizing it. Let me further state that Olympic lifters are highly skilled, dedicated to training and balance strength, flexibility, agility, and coordination.

SPECIFICITY OF TRAINING

If we spent time training our employees in the specifics of how to bend, lift, push, pull, sit, carry, and exercise until it was a built-in neurological habit; if we spent time training them how to move correctly to fit their job demands, and they kept themselves physically fit and with the help of proper ergonomics, I submit that we could stop back injuries at work and at home.

I am not a zealot! There are people, although few in number, who should never lift with their backs locked in. Some who have worked improperly 30 to 50 years at hard manual labor without backache, should not attempt to change their habits now. My observations of men who have been on railroad rail gangs for 35 years is that they have impeccable body coordination, use the locked in technique, and are free of back pain. If they did not have impeccable body coordination how could they survive 35 years of lifting and lugging rail that weighs 60 lbs. every 3 feet times 39 feet in length?

Certain anatomical or physiological problems can keep a person from being able to lift with this model. When confronted with this, I attempt to watch the person's work habits and, if convinced, leave them alone and pray they do not contract back pain.

STRENGTH TESTING

Strength testing alone has recently shown to be of questionable predictive value when testing a group population and not done for specific tasks.[11] Batti'e, et al., had a huge population of subjects tested for isometric strength in order to better understand the relationship of a person's strength and the occurrence of industrial back pain, a relationship not fully understood. It has been reported in the literature that isometric strength of two to four times that required for a job task may be a safety buffer for material handlers.

What Batti'e et al. discovered in this beautiful study was that general isometric lifting strength testing for workers who perform a wide variety of tasks is ineffective in identifying individuals at risk for industrial low back problems. As with any study, we could nitpick it. However, you can only look at so many variables in any study. The results should come as no surprise as people do not move or lift isometrically and there is no specificity of training.

With respect to lifting and industrial low back pain, we have yet to test my hypothesis; that is, specificity of training to the task, fitness, strength, endurance, and proper ergometrics in job site design will decrease back pain. All of these elements must work synergistically for best results.

Nothing we've discussed so far matters much if you implement a back school based on an ineffective model for lifting. The largest part of the back pain problem is related to improper lifting.[36]

MONKEY MODEL

As we have seen, lifting has been a long-standing, still raging controversy in medical circles. The historically accepted medical model has contended that man was never intended to walk upright — that, by walking around on our hind legs and lifting in that position, we are misusing our equipment. As proof, the experts point to gorillas, who do not have back pain. You never see a gorilla in the doctor's waiting room, do you? This is in direct conflict with the proceedings of the conference on idiopathic low back pain in 1980, Miami, Florida.[75]

Some of these experts have returned from travels to the Middle East with photographs of people working "in the squat", with their backs rounded out, and say: "These people don't complain of back pain."[21]

First, with regard to gorillas, I think the experts are reading the wrong book.[92] Also, I wonder how they can be so sure gorillas don't suffer from back pain. Second, there has never been, to my knowledge, a cross-cultural civilization study done to prove that any group is free of back pain.[74] In visiting foreign countries, including the Middle East, I've found the people overwhelmed with back injuries and back pain, just like us. A good friend of mine, Jitan Shah, a physical therapist from Texas, was born and raised in Bombay, India. He practiced there for nine years and told me 60% of his practice was devoted to low back pain, in a country that supposedly has no back pain.

This is a New Zealand sheep shearer.

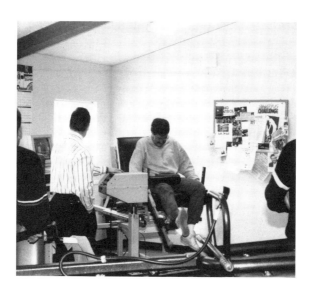

Back pain is international and universal. It is largely due to poorly understood mechanisms of injury, improper prevention and treatment. For 50 years, industry believed that people must lift with their legs, not their backs; that their legs are strong, their backs are weak. While experts keep teaching this, all evidence is to the contrary.

Computerized equipment for testing back and leg strength reveal that in "normal" people (those without back pain), back muscles can be two to four times stronger than the leg muscles.[66] In our tests of low back pain patients, we almost always find that they simply do not have the leg strength to lift according to the old or new model. In essence, they have to move their backs from total flexion (forward bending) to extension (standing upright) under load. I do not believe we were made for that.

Older biomechanical models insisted on describing the human spine as a cantilever system. To the contrary, according to Aspden,[8] the human spine can only accurately be described in geometric terms as an arch system, most stable with three inward curves. This model with three curves is the anatomical posture. Every first year medical, physical therapy, nursing and chiropractic student understands this as normal. Every anatomical text describes the anatomical position in the front of its text. Yet with more education, we forget normal and are led to believe erroneous theories.

LIFT WITH THE SPINE IN NORMAL POSITION

Since the beginning of my work with industry, I have opted for the anatomical position or normal posture as the correct approach to lifting. In other words, I teach lifters how to preserve the normal curves by keeping their spines in normal position. My choice is based on observations of power lifters and Olympic lifters. Some power lifters have back pain; most Olympic lifters do not. Given that as a fact, which do you want to emulate in lifting?

In dead lifting, the power lifter rounds his back when he reaches maximum ability to lift. Just like your typical worker, he will succumb and extend his legs, hips, knees and then his back. Because the back is extraordinarily strong, he uses it — and risks it. And, just like your worker, he is headed for back pain.

This logic probably conflicts with what you and your medical advisors have been led to believe. When talking to medical practitioners about this in seminars, I, the 150-pound pencil-neck geek, was nearly tarred and feathered a few times. That's when I started having Curt come along and demonstrate my points. Even now, some corporate medical doctors block their company from using my services because they just can't believe I'm right.

The old model was never anything but theory, given carved-in-rock credence purely through repetition. In the last ten years, the Olympic lifting model has been popularized largely through the American Back School's impact in industry, medical seminars, and the thousands of hospitals and clinics using the ABS system. Ten years ago, it was heresy; today, it is still viewed skeptically by many, but proving itself right every day.

NOT FOR EVERYONE

The Olympic model is applicable to about 80% of the work force, for both on-the-job and off-the-job situations. However, some workers are in such poor physical condition they need to get fit before converting to this model. Look at the photographic sequence that compares the old model to the Olympic model in work situations and beginning exercises designed to facilitate use of the Olympic model.

Back rounded out maximum, no muscle protection.

Although their backs are locked in, they are not following safe job
procedures at the worksite.

Curt and his fellow Olympic lifters present a most convincing argument: They live free of back pain. Experiences with patients in the clinic and industry have proven the validity of this model over and over again. It is my experience that a back school incorporating the Olympic lifting model can reduce lost-time back accidents by 20% to 40% in its first year. Using a five-year strategic plan along with education can eradicate back injury and its related costs in some industries.[10,84]

Researchers from Enoka[19] to Garhammer[28] have studied the mechanics and energy flow of Olympic-style weight lifting. The mechanics never suggest rounding the back. Pytel and Kamon wrote another excellent article to review strength as a predictor for safe maximal lifting.[55] They propose a model for predicting maximum dynamic lift in males and females and give a simple equation to realize that "magic number" using a portable isokinetic strength measuring device. McLaughlin, et al., in their study of the kinetics of squatting found that highly skilled power lifters kept their backs more erect than lesser skilled lifters.[48]

MECHANISM OF INJURY

Our hypothesis states that the trunk muscles (back and abdominal) should stabilize the spine in its normal posture while under load of lifting, pushing, pulling, etc. We get into trouble with lifting when we bend at the waist. We do not even have to lift anything. Sometimes it is just the act of bending at the waist that "throws the back out." Why? How can simply bending over cause such trouble?

As previously stated, the back muscles do not work when we are bent at the waist. Spinal ligaments and lower back spinal joints are stretched and the discs are under greater stress (stretch and pressure). As we re-extend, perhaps the muscles do not pull the joints correctly, causing them to get jammed. Furthermore, the muscles then "kick in" to keep you from further hurting your spinal joints. Thus, you have a back spasm. If bad enough, it could cause you to list, or shift your hips to one side.

WHAT REALLY HAPPENS

It all depends on who's camp you listen to and believe. Some believe all back injuries are due to discs, others blame the spinal joint, muscles, or spinal ligaments. My opinion is that improper movement causes the muscle, ligament, spinal joint, or disc to be sprained, strained, or injured. Whatever the mechanism, it is not known exactly. Certainly it is not worth arguing, but it is worth preventing.

By using the Olympic model to lift, push, or pull, you lock your back using your back and abdominal muscles and move the object with your hips and legs. This protects the ligaments, joints, and discs. It does so by using the only elements in your body that can protect you—your muscles! Most everyone has told you not to use your back because you will get hurt. Nothing could be further from the truth. When you use muscles they get bigger, stronger, flexible, and more durable. Why then, are you told not to use the only element in your back that is under your active control?

QUICK FIX OR LONG-TERM CURE?

Over the years, people have asked about the weight lifting belt.

In some industries, all the workers use them. My answer is this: If that is all it took, everyone would do it. Weight lifting belts help, especially in power lifting. However, I would only put a weight lifting belt on someone who is trained in proper technique for lifting, use of the belt and has good strength and flexibility. Contrary to popular belief, few people understand how to achieve proper dynamics between strength and flexibility.

Curt Cardine says the belt helps keep his back warm and reminds him to lock his back. It is not going to protect him if he lifts wrong because it is "a piece of dead cowhide". For most men, a weight lifting belt is probably not necessary and will not make them stronger. They should use their own muscles because muscles become stronger with use.

According to Dr. I.A. Kapandji,[82] the curves of the vertebral column increase the resistance to axial compression forces. Engineers have shown that the resistance of curved columns is directly proportional to the number of curves squared plus one. Applying that to the vertebral column, we have the formula $R = N^2 + 1$, in which R is resistance to compression and N is the number of curves in the spine. The medical lifting model has two curves, whereas the Olympic model maintains the three normal curves. Compare the equations:

$$\text{Medical model } R = 2^2 + 1 = 5$$
$$\text{Olympic model } R = 3^2 + 1 = 10$$

Mathematically, we can see that the spine has twice the resistance to compression in its normal position.

I will never believe it is good to shut off back muscles during manual labor. Some belt manufacturers would try to convince you to do so out of ignorance — both theirs and yours. But you are no longer ignorant and understand the importance of utilizing the back muscles to protect spinal bones, joints, ligaments, and discs. Unfortunately, I have gone to great lengths to show you how our thinking, beliefs, and traditions can keep us from understanding the truth.

CHAPTER NINE

Progress always involves risk; you can't steal second base and keep your foot on first.

Frederick Wilcox

THE RIGHT COMMITMENT GETS RESULTS

We live in an "instant society." Can you remember turning on the TV, then waiting for it to "warm up?" Today, it's instant coffee, instant tea, instant meals, instant just-about-everything. We might very well name this The Age of Impatience. Unfortunately, there's no instant fix for back pain problems. One round of back school classes will have a temporary effect, but no long-term, lasting results.

When making a commitment to get back problems under control, you must stand ready to "put your money where your mouth is" because it will be a line item on your budget for the next several years. It's up to you to determine whether it is an expense or an investment. If you properly view it as an investment, you must successfully communicate that commitment to your associates, Board of Directors, stockholders, and employees.

Doing it right will mean bringing in consultants, training trainers, and taking workers off the job to attend classes. It means forming ergonomic committees and making job site changes. Doing it right means commitment to a comprehensive, continuing program. I suggest five-year plans to my clients.

All of this costs money. For example, given 10,000 employees, you may incur a first year cost for training trainers, conducting classes, providing educational materials to all employees, etc. from $30,000 to $150,000. That's not pocket change! On the other hand, considering current total costs of back pain and back injuries, it's not a very large amount at all. How small of a percentage of decreased lost-time back accidents and related costs do you really need to break even on your investment? If you accurately and honestly make these calculations, you'll be amazed at how little it takes to profit from this investment. Just one, two, or three lost-time back accidents prevented each year will probably return your investment. Our clients typically report annual return on investments of 200% to 500% (see chapter seven).

Compared to other aspects of your business in which you routinely invest similar amounts, (probably for a much lower return on investment), back safety education offers a measurable return on your money. Companies spend money every day on ad campaigns whose results can't be measured. And what about that piece of equipment that will take 20 years to justify its cost?

If you sincerely want to get back problems under control, you will have to make an appropriate economic and time commitment. There is also a nonfinancial commitment to be made, and it has to come from the very top down through the ranks. Top management must make a commitment of attention and of evidenced priority. You need to get involved in the statistics, in taking part in the classes, in touring the job sites, and in supporting the ergonomic committees. Trainers and supervisors must be given time during the work week to meet, analyze, plan, implement ergonomic changes and conduct classes without feeling they are getting behind in their other work. In other words, to be blunt, you can't just give a budget and some lip service to this — you must give real, visible commitment in action.

TRAINING MAKES DOLLARS AND CENTS

In their article on back injury prevention training, Toner, et al.,[71] give many statistics with respect to decreasing injuries and costs. The bottom line was that for every dollar invested in training, $6.50 was saved.

Now that's an excellent return on investment! The following data is from other companies we have served. Let me point out that we had nothing to do with its compilation. We were, and still are, at the mercy of the industry to provide us with proper data. I submit the following examples:

CHESSIE SYSTEM — APRIL 1984

	Manhours	Lost Time Accidents
September 1980-81	11,000,000	55%
September 1981-82	12,000,000	11%
September 1982-83	9,000,000	26%

Training seminars were conducted September 1981 through April 1982 for the West Virginia Division and systems operations at Raceland Car Shops and Russell Locomotive Roundhouse. No follow-up training was pursued past April 1982, or in 1983 by Chessie System. (The company was to have changed their method of injury data collection in 1982, which was supposed to help the number of LTA in 1981-82, according to a company official.)

MISSISSIPPI POWER COMPANY

1982-1983
7 Years Post-training
No Lost Time Accidents - 1600 Employees

Three years prior to training, Mississippi Power Company had experienced 171 lost-time days due to back injury. Each class included about 25 participants and lasted two hours. All training was done by Mike Rogers, P.T., MNFF.

WESTMORELAND COAL COMPANY

1)	Severity Rate	1982	88
		1983	100
		1984	77
2)	Man Hours Worked	1982	2,987,447
		1983	2,918,539
		1984	2,962,943
3)	Lost Time Workshifts	1982	1,310
	Due to Back Injury	1983	1,466
		1984	1,136
4)	Difference between '83 and '84		
		1,466	
		−1,136	
		330	
5)	Estimated Savings	$161,130	

Three-hour training programs teaching prevention of back injuries were conducted as part of a 1983 annual retraining. Class size varied between 10 to 35 participants on all shifts and over 1500 employees were trained from January until October. No follow-up training for back injury reduction was given in 1984 annual retraining. These statistics were derived from Westmoreland data. They may be more conservative than they should due to the impossibility of pulling all miners' files for pay scales.

BACK INJURY DATA - THE CHELSEA PROGRAM

Year	Back Inj.	Mean Pop.	Cases Open	Days Lost	Comp. to dte	Med. to dte	Comp./ Med. tot.
1977	32	580	0	1082	$211,706	$31,254	$242,960
1978	13	404	1	804	$140,881	$20,820	$161,701
1979	6	307	1	937	$187,040	$28,663	$215,747
1980	11	359	2	378	$39,865	$7,945	$47,810
1981	5	387	0	35	$1,219	$60	$1,279

PEABODY COAL COMPANY

Twenty-seven trained in Train The Trainer program in 1985. In 1986, ONE DIVISION ALONE had an estimated savings of $500,000 in worker's compensation. If the other four divisions did half as well [($250,000 × 4) + 500,000 = $1.5 million saved], then Peabody had a tremendous return on investment!

MARROWBONE DEVELOPMENT COMPANY

In January 1988, the company intensified their training program and implemented a ten-minute stretch and flex program. Training centers were installed at mine sites. Marrowbone Development[10] went 22 consecutive months without one lost-time accident due to back pain.

CLINCHFIELD COAL COMPANY

	1982	1983	1984	1985	1986	1987	1988	1989
Underground	34%	31%	58%	23%	23%	15%	22%	16%
Surface	25%	36%	31%	20%	23%	18%	21%	21%

This report on LTA's due to back pain gives us wonderful insight into the nature of ongoing training and its effects. By 1984, Clinchfield Coal was concerned that back injuries were getting out of control and decided they must do something about it. Over 600 underground miners were trained in 1985. During that time, two employees (one company, one union) were trained to become trainers. Those two trainers attended every moment of training, underground work followup and job site inspection. I am confident they are the two best trained trainers I've ever worked with. After September 1985, the two trainers trained an additional 2,000 + miners. With continued followup efforts, we see results in the second, third and fourth years after initial training plus a decrease in the severity of injury. All of this while the UMWA was selectively striking against the company!

To put it boldly, **if coal mine companies can reap these results, you can, too**! I am not a hero. I had no desire to descend into the 25+ mines I had been in and spend three and a half years of my life working under mountains. However, I concluded that if we could prove that back injury prevention training can work in coal mines, it could work anywhere. Is that too bold for you? We are talking about one of the harshest working environments in the world; yet, they have reported excellent results.

MUELLER PIPE & GAS COMPANY

Mueller is another tough industry. In fact, I walked out of their facility in Albertsville, Alabama where fire hydrants are manufactured, and said I would rather work in a coal mine. But with ergonomic changes and commitment to education, they have done a wonderful job in getting their back injuries under control. One year after training foundry workers and implementing ergonomic job site changes, I asked Layman Ferguson what he thought the program had done for him, his workers, and his company. He stopped dead in his tracks, looked at me with a very serious sober look and said, "It was worth its weight in gold." That is how we came to refer to the Train The Trainer program as worth its weight in gold.

CHAPTER TEN

They can because they think they can.

Virgil

PRACTICAL ERGONOMICS: SIMPLE WORKPLACE CHANGES THAT MAKE SENSE

Webster's Dictionary defines *ergonomics* as 'biotechnology.' However, my definition is simpler: creating a work environment where man and machinery work in harmony, without putting each other at risk, for optimum productivity and longevity. Another way to think of it is simply to have the job site facilitate safety.

Some ergonomists frighten industries. The idea that huge sums of money must be spent in complete make-overs of job sites, equipment and machinery is intimidating and discouraging. I prefer to start with a "one step at a time" idea. And I've found that even small ergonomic changes can yield very big dividends. It has become painfully obvious that no matter how well we train people to prevent injury, if we put them back to job sites that put them at risk for injury, we just spin our wheels.

Let me give you a simple example of minimal but valuable ergonomic change. The palletizing and depalletizing of bags of grain, concrete, or other materials is a job function that typically

creates significant risk for injury. With a full pallet on the floor, lifting from the top four rows is easy. But when workers reach the bottom rows, they almost certainly will bend at the waist to pick up the bags — one of the worst lifting movements possible!

Depending on the position of the pallet in relationship to work space, the individual in this job will rotate to the right over and over again until he develops pain in his right leg or the right side of his back. He then may switch everything around and rotate to the left until he wipes out the left leg or left side. Eventually the body just gives out. You can train the person in this job for three hours, three days, or three months and it will not matter much if he does not acquire safe movement patterns as rote habit. Obviously, if we can change the work site to eliminate the hazard, we can standardize a safe work site for all employees working that job.

ERGONOMICS AT WORK

The easiest way to resolve this situation is with a device commonly called a "palletizer," which picks up the pallet and maintains it at waist level. A forklift can be used to do the same thing.

However, my observation is that forklifts are too busy to be used statically. There never seems to be enough of them for this kind of work. Compared to the cost of one back injury, a palletizer is a bargain.

Here's another example: A steel mill had 40-pound bags of flux that had to be lifted out of bins. Lifting from the top of the bin was, of course, no problem. But lifting from the bottom of the bin not only required a height of about six feet, but also bending at the waist. Lifting that weight while bent at the waist through that long lever arm of the spine, then extending with the weight of the bag and the torso hundreds of times each week is going to cause incredible wear and tear on the back. When you are completely bent over at the waist, the back muscles are completely "shut off"[21,27,29] so that there is no protection for the lower back. Again, we can train the worker for hours, weeks, or months; but when we return him to a job that defeats his body's natural protection, it will only be a question of time before he is injured.

In this instance, the company's employee ergonomics committee discovered a very cheap, easy solution. Instead of building a bin three feet high, the carpenter shop now builds it only two feet high, placing it on angle irons so it tilts forward slightly and is between shoulder and waist level. Workers can pick up all the bags from a safe position, turn, pivot, and walk four steps before letting go of the bags. If they did not have to walk four steps, they would pivot the torso on a stationary pelvis and throw the bags

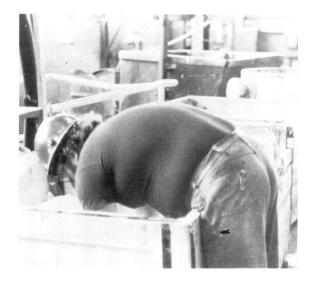

into a vat. This would place the worker at high risk because people never pivot correctly, although they know they should.

Why don't they? It is not an ingrained, rote movement pattern. This is the same as a baseball pitcher throwing 250,000 pitches each year. He does not think of where he is going to place the ball, it is done by rote. But under pressure in the World Series, he starts to think about placing the ball, steps out of his rote movement pattern and loses all control.

Opportunities for ergonomic changes are in plants and facilities everywhere. If approached with excitement and anticipation of permanently creating safe working environments, the ergonomics program can help workers now as well as generations of employees to come.

OFF THE JOB INJURIES

Ideally, training should change habits both on and off the job. Here are two situations in which the individual can take care of himself:

Bending at the waist to load or unload the dishwasher may be a benign experience for most of us. To back pain sufferers, it can be another activity that keeps them in chronic pain. Here the model bends at the waist, hips and knees and leans on one leg with his left hand to support some of his torso weight.

After being bent over working on the engine of a car, his back became stiff. Simple standing back bend exercises, popularized by Robin McKenzie, PT, can help alleviate that stiffness.

Does this job look familiar? Bags weighing 50-60 lbs. must be lifted out of this crate. It's pretty easy when they are at the top, but what about the bottom? How would you change this job to make it safer?

How about here? With the back rounded out you are at risk! Why? The back muscles cannot work to protect you. (Possible solutions on page 164.)

ON THE JOB INJURIES

Here is a picture of the steel bar plates (estimated weight,120 lbs.) used in our NIOSH lifting example. What's wrong with this picture?

Here are the benches to which they must be lifted. What do you think can be done to minimize the risk of lifting this heavy piece of metal?

(Possible solutions in appendix.)

This picture shows where the typical motor (estimated 150 to 175 lbs.) was placed. Because many injuries have taken place repairing or replacing these motors, they have been extended for easy accessibility.

Note how easy it is now to get to the motors. That is exactly what ergonomics must do — change dangerous conditions to safe ones.

ON THE JOB INJURIES

How does this man look? That's right, he's an accident waiting to happen. Training will help change poor habits like this to better ones. However, what measures would you take to ensure complete safety with this shoveling job?

This man is lifting bundles of roof bolts estimated to weigh 60 to 70 lbs. In this posture, is he protected? What would be a better movement pattern?

This man is sitting in a continuous miner. This one is new and ready for service. The model is a 6′ 2″, 210 pound man who, if he were a miner, would be expected to sit in this seat 5.5 to 6.5 hours each day. Notice the seat! It is slab steel with a covering of hard rubber. The deck on which he sits vibrates violently all the time he

is on it. His knees are much higher than his hips, causing his back to be rounded. Therefore, his back muscles cannot help absorb the shocks and vibration transmitted to it. The stress falls on his vertebrae, discs, joints, and nerves.

Vibration has been studied at the University of Vermont and shown to have detrimental effects on the human spine. How long would you last working like this? How can mine operators expect their people to work safely like this? Why do designers make equipment like this when they know it injures the operators? We have much to do to correct these situations.

CREATING AN ERGONOMICS COMMITTEE

Early in my experience with ergonomics, I worked closely in a steel mill with a very knowledgeable Advanced Safety Engineer Charlie Simpson. Together we assembled a team composed of:

1. plant physician
2. plant nurse
3. upper management representative
4. union representative
5. health and safety manager
6. personnel manager
7. shop supervisor
8. an engineer
9. a consultant, such as the physical therapist assisting with back education
10. individuals representing different job functions and sites

Over a period of time, in concert with training about 1,500 workers, the team made changes in about 30 job sites. Most required very small expenditures. My colleague, Dennis Hart, Ph.D., P.T., has gotten very involved in ergonomics and flatly states that most job sites can be successfully modified for under $500.

In addition to the direct benefit of injury risk reduction, there is an equally important though less visible benefit of forming a team and making changes. When a company begins a comprehensive training program, there are always a few loud, negative voices criticizing the program and pointing to dangerous job sites as proof that the company is "all talk, no action". The formation of an ergonomics committee and its implementation of some changes silences such loud-mouths and evidences the true commitment of the company. If cost is of real concern, the ground rules can be pre-established. The position of management might be to fund up to *x* number of job site changes over *y* number of months, each costing no more than *z* dollars, then take *a* number of months to evaluate the results before funding the next round of changes. I see nothing wrong with starting small, making the effort prove its worth and earn its keep. The important thing is to get started.

Dr. Stover Snook, an ergonomist with Liberty Mutual Insurance Company, has opined that ergonomics alone can cut lost time back accidents by as much as 33%.[67] And, although I am a staunch advocate of training, I think ergonomics changes in the workplace are at least as important in some cases, possibly even more important than education.

Since the introduction of the NIOSH work practices manual in 1981, the algebraic formula for describing action and maximal permissible limits have been used to describe safe lifting. During training, I discovered calculators, slide rules and computer programs are available to examine lifting tasks and their compression on the low back. As part of the ergonomics program, this is going to be a vital tool in job site screening or assessment.

The action limit describes that over 99% of men and 75% of women could lift loads described by this action limit (AL). The maximum permissible limit (MPL) is three times the AL, and only 25% of men and less than 1% of women workers have the strength to work above that level.

Four factors modify the formula:

$$AL = 90(6/H)(1\text{-}01[v - 30])(.7 + 3/D)(1 - F/FMAX)$$

1. H = the horizontal factor — distance of load from midplane of body, usually ankle bones
2. V = the vertical factor — distance from the hands to the ground when beginning the lift
3. D = distance load is lifted
4. F = number of times load is lifted per minute

Admittedly, there are restrictions with this formula in that most people don't lift from the floor to table top. However, it is the only formula to date that may give us a clue as to the need for job site redesign to help decrease the risk of occupational low back pain.

REAL LIFE CALCULATIONS

The following examples are taken from a recent job site evaluation in which the foreman thought the steel bars weighed 30 lbs. In fact, they were 90 to 120 lbs., revealing that few people correctly calculate weights they lift. Each example shows the compressive forces on L5S1 for 30 through 120 lbs. This tool can help you and your ergonomics team understand that certain job sites must have changes in order to be safer.

BIODEX LIFT WORK SCREEN TOOL

Individual Lifting:		Load Lifted: 30 lbs.		**Lift Position**
Height:	72			
Weight:	180			
Sex:	M			

Parameters:		**Angles:**	
Hand Position:	3	Elbow:	15
Ankle Position:	16	Shoulder:	15
# Repetitions:	3	Hip:	140
Back Position:	BBI	Knee:	130
Attachment:	Between Knee	Ankle:	35
Test Position:	FULL SQUAT		

Joint Performance Index (%)

		Elbow:	14
L5/S1 Compression: 791.44 lbs.		Shoulder:	3
% MPL:	55	L5/S1:	26
% AL:	103.59	Hip:	35
		Knee:	47
L5/S1 Shear: 71.18 lbs.		Ankle:	59

This shows starting position of the lift. Note the joint angles. I simply measured each body part, recorded it and later fed it into our computer. The L5/S1 compression is about 50% of maximal permissible standards and is over 100% of action limit.

BIODEX LIFT WORK SCREEN TOOL

Individual Lifting: Load Lifted: 60 lbs.   **Lift Position**

Height:	72
Weight:	180
Sex:	M

Parameters: **Angles:**

Parameters		Angles	
Hand Position:	3	Elbow:	15
Ankle Position:	16	Shoulder:	50
# Repetitions:	3	Hip:	140
Back Position:	BBI	Knee:	130
Attachment:	Between Knee	Ankle:	35
Test Position:	FULL SQUAT		

Joint Performance Index (%)

		Elbow:	27
L5/S1 Compression:	1054.35 lbs.	Shoulder:	5
% MPL:	73.27	L5/S1:	32
% AL:	138	Hip:	44
		Knee:	56
L5/S1 Shear:	90.27 lbs.	Ankle:	68

Here we see the same style lift with a heavier object. Note the compression on L5/S1. Few people can work for sustained periods of time at this level, as is seen with MPL.

BIODEX LIFT WORK SCREEN TOOL

Individual Lifting: Load Lifted: 90 lbs.   **Lift Position**

Height:	72
Weight:	180
Sex:	M

Parameters: **Angles:**

Parameters		Angles	
Hand Position:	3	Elbow:	15
Ankle Position:	16	Shoulder:	50
# Repetitions:	3	Hip:	140
Back Position:	BBI	Knee:	130
Attachment:	Between Knee	Ankle:	35
Test Position:	FULL SQUAT		

Joint Performance Index (%)

		Elbow:	40
L5/S1 Compression:	1317.26 lbs.	Shoulder:	7
% MPL:	91.54	L5/S1:	39
% AL:	172.42	Hip:	53
		Knee:	65
L5/S1 Shear:	109.35 lbs.	Ankle:	77

In this example we are at 92% of maximum permissible limits for a heavy infrequent lift. Most people (male) cannot lift over 35 lbs. repetitively using the NIOSH guidelines. Certainly lifting heavy weight like this frequently excludes much of the population. Remember only 25% of male workers and less than 1% of female workers have the strength that allows them to work safely at these levels.

BIODEX LIFT WORK SCREEN TOOL

Individual Lifting:		Load Lifted: 120 lbs.	**Lift Position**
Height:	72		
Weight:	180		
Sex:	M		

Parameters:		Angles:	
Hand Position:	3	Elbow:	15
Ankle Position:	16	Shoulder:	50
# Repetitions:	3	Hip:	140
Back Position:	BBI	Knee:	130
Attachment:	Between Knee	Ankle:	35
Test Position:	FULL SQUAT		

Joint Performance Index (%)

		Elbow:	53
L5/S1 Compression: 1580.17 lbs.		Shoulder:	10
% MPL:	109.81	L5/S1:	45
% AL:	206.83	Hip:	62
		Knee:	74
L5/S1 Shear: 128.43 lbs.		Ankle:	86

Approximate weight of the steel bar plates in bottom picture, page 85, is 120 lbs. Obviously, we have exceeded MPL.

I used this tool to catch the eye of the company's safety officer. He was up against a brick wall in the shape of a foreman. The foreman said nothing was wrong with the way the job was performed, that the load only weighed 30 lbs. He said he could pick it up from the floor with one hand! It took the author to his maximal dead lift limit. Our search of the records revealed four people were off this job site for two years or longer due to back pain. Obviously, there was a problem. But this supervisor foreman was adamant, if not downright abusive, to the safety director and myself.

The graphs I have presented refute the foreman's objections. They allow upper management, ergonomics committees or open-minded supervisors information that can enable them to act on job site redesign for the safety of all employees.

I must admit being uneasy using the NIOSH guidelines. Intuitively, something just seems wrong with the formula and its outcomes. To me, the outcomes measure a light load of 30 to 35 lbs. that can be lifted frequently within safe NIOSH limits. Having worked at many real life manual labor jobs, I know it is not a realistic load limit.

Also, it measures movement only for a straight saggital plane lift in which the weight is lifted vertically and perpendicular to the ground without turning the torso or moving the feet. Straight saggital plane lifts rarely occur in the real world. Robots move that way, not people!

I do not know who has scientifically validated or found this to be reliable in the real world — that being your individual situations, but it has been carved in stone without proper validation. That is frightening, just as the medical model for lifting is frightening. However, it is the only lifting formula we have. And with respect to its authors, I know it is the product of much work. So, as with all changes, we have to start somewhere.

Observations show that people do not lift with straight saggital movements but move dynamically through three planes:

1   bending — forward and backward
2   rotation — twisting
3   side bending — almost always coupled with rotation

Therefore, we must seek methods that better relate to the forces that dynamic human beings put on their bodies.

Other formulas describe safe lifting by using isometric lifting models. We must understand that lifting, pushing, and pulling are isometric only until the object being lifted is in motion (a millisecond), then it becomes a dynamic movement. By having a better understanding of the Olympic weight lifting technique, we can see this principle in action and measure dynamic movements.

CHAPTER ELEVEN

Well done is better than well said.

Benjamin Franklin

ADVANCED TRAINING

The ABS Advanced Training has a prerequisite; the trainer must have completed our Foundations of Training course and have had field-proven experience using our concepts within their industry. Exemplary communication skills in teaching through lecture, demonstrations and constructive critiques are a must. Safety trainers need advanced training on back education. The fundamentals are important, but then as a review and refresher, I would recommend advanced training. This training also helps the trainer to become more confident in skills and presentation. It can also help the trainer play an even greater role in reducing the cost of back injuries within the company. Let me share what the ABS Advanced Training consists of to you a better idea of how valuable it is.

The advanced trainer also must be physically able to exhibit the advanced exercises of bending, lifting, carrying, and pushing as well as understand how to organize the ten-minute flex and stretch programs.

A large segment of time spent in training is to help the trainers and their company stay on track with their five-year plan. We

break into small groups, allowing time for brainstorming and dealing with the issues that personalize each back attack plan. Once the components of the plan are identified from the generic plan, a timetable is set and the plan can begin its second phase.

For years two to five, we have slide and video programs for lecture series, posters, exercise programs, safety talks, and even

company newsletters. Each facet of the program implementation is covered explicitly.

A payroll deduction plan, available for the employees and their families, is offered both in lecture and during training. This plan has home back and neck care systems, including videos, for ending back or neck pain and can be provided the following ways:

1. 100% paid by company
2. 75% paid by company and 25% paid by the employee
3. 50% paid by company and 50% paid by the employee
4. 100% paid by employee with a small deduction per pay period

Upper management is encouraged to fully reimburse employees for their expenses if they work one year without a lost-time back or neck problem.

The kits are fun, easy, safe and designed to help you get and keep your LTA's and dollars under control.

The payroll deduction plan is offered so that employees and their families will have something in their libraries to watch, study, and practice until it is rote habit. This is a powerful way to augment training and reach employee dependents. If you are self-insured and responsible for those dependents, it behooves you to make sure they are also not at risk.

Upon implementing a payroll deduction plan, fully explain the program — why it is important and how it works — to employees and unions. If they understand the purpose of the payroll deduction plan, they are more likely to accept it without problems.

The pain questionnaire, or **Functional Assessment Inventory** (FAI, Dr. Rose, et al.)[59] is reviewed and, if possible, tailored to each industry in order to be used to its fullest potential. The directions for properly administering the FAI are explained. Meetings can be more meaningful if you know in advance where back problems are, and the FAI is designed for that purpose.

The other integral part of advanced training is teaching trainers how to administer the risk factor analysis for low back trouble (LBT). Using six questions and ten measurements from the literature, the form is designed to reveal an individual's musculoskeletal factors that may indicate the potential for back trouble. One measurement, Shober's test, is of questionable value because of poor agreement concerning reliability of measurements among testers. Although not 100% accurate, it can tell you how the low back is or is not moving. This instrument is to be used solely to identify those at risk for low back trouble or recurrence. It is not to be used to qualify or disqualify for employment.

Once an existing employee who has five or more musculoskeletal risk factors has been identified, your company policy can dictate what should happen next. Obviously, you could do nothing. Ideally, a physical therapist should show the individual how to resolve the risk factors and the worker can be held accountable for periodic rechecks to ensure a change takes place and becomes permanent.

FIGURE 1. RISK FACTORS QUESTIONNAIRE.

Name: _____   Age: _____   Date: _____

1. Are you involved in a regular routine   ___Yes   ___No
    exercise program?
    If yes, how many times a week?   _____
    How many minutes at a time?   _____

2. Are you a cigarette smoker?   ___Yes   ___No
    If yes, how many packs per day?   _____

3. Have you been through a back education
   program?   ___Yes   ___No
   If yes, how long ago was it?   _____

4. Does your job involve heavy regular lifting? ___Yes   ___No
   If yes, how heavy is the heaviest
   weight you handle?   _____
   Times per hour?   _____

5. Does your job involve lots of vibration?   ___Yes   ___No
   If yes, what is your job? (circle one)
   Driving truck, driving car for a living, jackhammer,
   land haul vehicle, bulldozer, other   _____

6. Have you ever had back pain before?   ___Yes   ___No
   If yes, missed how much work?   _____

Measurements

                                                  Circle One
7. Posture analysis   Normal   Swayback
                                        Flatback   Lordotic

8. Forward Bending   _____   Back tight
                                        _____   Hamstring
                                                    tight

9. Hamstring II Length   90-0 degrees _____

                                        Key for grading 0 - 5

10. Curl sit-up legs straight   grade   _____
    Optional leg fallout

11. Lower abdominal test   grade   _____

12. Prone press-ups   0% Motion   _____

13. Prone trunk extension   grade   _____

14. Schober's test   _____ _ cm
    15-18 cm = hypomobile
    above 23 = hypermobile

15. Proper lifting model demonstration ____ Pass   ____ Fail

16. Deep squat - feet flat, hips
    below knees   ____ Pos.   ____ Neg.

                                        Circle One
17. Results = positive _____ of _____   Risk Factors None   Minimal (1-4)
                                        Moderate (4-8)   Severe (8+)

By using this form, we can have an idea of who may be at risk for LBT. Each participant spends 15 minutes being measured. The six questions are answered prior to measurement. Advanced trainers need a total of $5 to $6 of equipment to administer this test. At the end, risk factors are added up, scored, and categorized as zero, slight, moderate, or severe risk. Corporate policy must dictate what action is taken with the moderate-severe group. My advice is to refer immediately to one of the PPN (see Chapter 13) teams and let them intervene with proper flexibility, strengthening, coordination training or even other therapies to rule out the chance of a serious problem.

TRAINERS AS SCREENERS

Trainers are screeners, not treaters. To conduct this program properly, you must have a health care support team. It is my hypothesis that you will pay more for those employees who are out of shape. They will suffer more back pain if they do not know how to change their risk factors. The program is similar to screening for cardiac risk factors, which has succeeded very well.

An important lesson was learned when the program was implemented at a major oil company. Unfortunately, the program was voluntary and was hampered by poor communication with management. However, 70 to 80 people, out of 200 who participated, were well pleased. As expected, some followed our suggestions

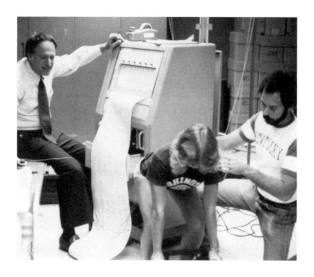

and in a two-month check (enough time to allow for stretching tight muscles and ligaments and strengthening weak muscles) had resolved their risk factors. Some were still working on them, others were struggling, and some just plain quit.

YOUTH AT RISK — AND THEY DON'T EVEN KNOW IT

The extreme tightness of hamstring muscles and low back hyperflexibility in participants ages 21 to 25 shocked me. Along with those problems is extreme weakness of abdominal, back and leg muscles. They had just started their careers, with multiple jobs that required frequent bending and moderate to heavy lifting, and were already at risk!

I concluded there was not a higher incidence of LBT due to the youth of the population of employees. But let's think for a moment about where these young people will be in ten years? Will they be on your company's disability payroll, in pain, or forced to change occupations?

Many times we give little to no regard to what and how we do a job. But what we do today may dictate what we do or don't do five to ten years from now. The movement patterns and habits for accomplishing job tasks we acquire today must be proper so as not to put our bodies at risk. The good news for most of us is you can get by with most anything until you are 20 to 30 years old. At that time, the body starts to change and backaches may become more frequent, if not constant.

Again, I do not believe back pain has to be a cost of doing business. I believe the risk factors of the employees, job sites, and other factors must be evaluated, identified, and for those at risk, rectified. If you do not want your trainers doing that, it is alright. A health care practitioner such as a PT or PTA could be of immeasurable help in assisting with this part of your five-year plan.

WE SHOULD BE EXCITED

It has been said and written in the literature *ad nauseam* that only 1 to 2% percent of all back pain sufferers need surgery. That should excite every one of us because not only do the 98 to 99% need us, what about the 1 to 2% after surgery? If we do not reach them and change their poor movement patterns, muscle tightness, weaknesses, and job site design, what's to keep them from having another surgery, injury, or permanent disability? You see, the great thing about us as trainers is that everyone needs this information.

As trainers, we must share fresh and vital information that is immediately applicable. Our society must change its attitude that doctors will fix our problems and always take care of us. Thank God for doctors, but many times a person presents himself to a doctor when it is too late for the doctor to completely cure this person of permanent pain or disability.

I will believe until the day I die that back injuries and pain do not have to be a cost of doing business. However, I will believe that dollars spent in the prevention of back injury are a cost of doing business and always will be. The preventive dollar spent is always wiser and much less than cleaning up after the accident. With the soaring costs of medicine, surgery, hospital stays, and therapies, it has become a must for preventing problems rather than treating them.

Recently, Mr. Jerry Davis, president of CSX Transportation, Inc., spoke to the Chamber of Commerce in Ashland, Ky. Chessie System, now a part of CSX, is the first industry I served. His address could have knocked me off my chair. Mr. Davis stated that bottom-line profitability of any company *must* be connected with the safety of the work force!

My report to Chessie System ten years ago read the same as Mr. Davis's speech. You can not just preach safety — you must do it, and do it daily. If it is just talk, it is of *no* value, for the words are empty. It is time that safety, personal and environmental, be the top priority of every company on earth. With top management committed, like Mr. Davis, CSX will do well in the future. Their safety record already reflects this and productivity can only get better.

ADVANCED FIVE-YEAR PLAN: KEEPING TRAINING ON TRACK

Statistics have proven over and over again that the companies who have the best results in back injury reduction were the ones who implemented a comprehensive, long-term plan. Their back programs were more than just crisis intervention.

The five-year plan was put together in order to coordinate the implementation of a back injury prevention program with the involvement of your employees, supervisors, management and outside consultants, medical, or other. Its specific intent is to give you a plan to put in place to keep your efforts alive weekly throughout 60 months. Obviously, you must alter it to fit your specific needs. You will find years four and five to be repetitious. By year four, you should be on track and managing back trouble easier.

By executing your plan over five years, your data will help you see trends in injuries. For example, if LTA's due to back pain surface in a certain area, you can immediately administer change and bring the problem under control. It will pay handsome dividends to control your data.

Just a reminder — literature abounds with informative articles about occupational back injuries. Some startling facts:

1. If off work for six months due to back pain, a worker has a 50% chance of returning to work.[89]
2. If off work for 12 months because of back pain, he has a 25% chance of returning.[89]
3. If off two years due to back pain, a worker has almost no chance of returning.
4. Low back pain may only be 20% of compensable injuries but it represents 40 to 80% of costs.
5. People are not being educated about the problem. Deyo, et al.,[85] found that patients want better explanations of their symptoms.
6. Once a person has had a back injury, he is likely to have recurrences in the next two or three years.
7. We are spending between $50 and $100 billion each year due to back and neck pain.[4,25,77]

In order to make the implementation of a five-year plan simple and easy, we have developed a complete resource package. In other words, you can have a complete on-the-shelf package and by following the time table and using all the elements of the five-year package, you will be on your way to controlling lost-time accidents due to back injury.

Components of Five-Year Kit

Videos

1. Flex and Stretch - 6 to 12 weeks of exercises. Length of each routine complete with leader's cassette is 6 to 7 minutes.
2. Safety Talks - 5 to 10 minutes of timely topics. Twelve segmented talks complete with leader's cassette so that safety talks can be conducted at least once each month.
3. Keeping Trainers on Track - to inspire your trainers and keep them enthusiastic about their jobs
4. Payroll Deduct Programs - "End Back Pain Forever" and "End Neck Pain Now" target dependents of your employees as well as your work force, a complete program in itself that can be used over and over again at home.
5. Fighting Back Video - fifteen-minute video using proven principles to control back injuries.

Manuals

Give step by step instructions on how to conduct:

   a. supervisor safety talks
   b. use the slide programs
   c. keep the training on track
   d. flex and stretch

Slides

Five packages

   a. initial 2 to 3 hour program
   b. two 30-minute programs
   c. two 60-minute programs

Posters

You should purchase generic posters, then personalize using your own employees.

They should show:

a. lifting
b. pushing/pulling
c. sitting
d. exercise
e. work site dirty/hazardous
f. work site clean/safe
g. work site after ergonomic change

Your poster campaign could even have a main character or theme. Change posters at least every two weeks.

Books

a. *Back Book* - This can be personalized with your name, company logo, and can include a company safety message.
b. *Neck Book* - This too can be personalized for your company.

These books can be sent to your employees' homes with a simple three-paragraph letter reminding them of the benefits they received through back safety training. This letter should encourage them to use the book as a follow-up and to share information with their families, friends, and neighbors. The books should be sent one month post-training. They can also be used in conjunction with a payroll deduct plan, as they come with the videos "End Back Pain Forever" and "End Neck Pain Now."

Super Supports

The supports can be individually controlled for dynamic low back support. They're great for safety awards or can be purchased on payroll deduct plan. The super low back supports are also available with your name, logo, and a safety message. Providing these supports gives employees something they can use to prevent back pain at home, in the car, or anywhere. Plus, they are a constant reminder of the training they have just received.

Become a Member of a Preferred Provider Network (PPN)

The PPN links you with trained health care practitioners in your area who follow sound treatment and prevention philosophies and use them daily in their practices. If and when people get hurt, they deserve to be treated like human beings, with timely and affordable care. A PPN should have a list of members who meet the criteria for ensuring that type of care. All doctors are not alike. Not all can treat back pain. Nor can all PT's treat back pain. That is why we have specialties within both fields.

For most of the past 11 years that I've worked with industry, I have admittedly taken an unstructured approach to the prevention and control of back injuries at work — providing train-the-trainer seminars, all-employee seminars, books, on-site ergonomic inspections, and other consulting on a requested basis. Some companies went at it full blast for a few months, and some for a year. Some got great results from their first efforts, stopped doing anything, said the problems gradually climbed back to prominence, and then attacked them again. In the last several years, realizing the overall inadequacy of this approach, I've devoted a great deal of energy to developing a structured, long-term, committed approach to eradicating back injuries and even back pain from a given workplace.

There is no doubt that companies that commit to long-term, continuous efforts get the best results. With this obvious truth staring me in the face, I sat down and formulated a five-year back attack plan. The plan incorporates all the things companies most successful at solving the problems have been doing in continued follow-up to initial training of their supervisors and employees.

Why five years? The costly problems of back pain and back injuries occurred and built up over years, not months, and it logically must take years to remove them from a work environment completely. If you and your people could drop everything else they're doing and focus on nothing but this, we might talk in terms of a month or two. But, obviously, there are many other priorities vying for their time and attention. There is also the matter of conditioning, of habits hard to break, and breaking them takes repetition.

If you're going to implement an effective program, make the commitment to stick with it at least five years. Then make sure that you use a variety of materials and presentations so that your trainers and employees don't get bored. Let's face it, showing the same old slide program twice a year for five years is going to get old. Keep the program fresh, exciting, and fun!

Year One: Massive Awareness

During the first year of the five-year back attack plan, the single most important objective you can set and achieve is massive awareness of the existence, size, scope, and cost of the problems, as well as awareness of solutions that exist and can be successfully integrated into your work environments.

Have you ever bought a new car, thinking it was a pretty uncommon model but after having it, suddenly noticed there were a lot of them on the road? Before owning that particular car, there was no reason for you to be aware of its popularity. Because you own one, you automatically became aware of its popularity. Just creating a new level of awareness linked to self-interest will go a long way toward reducing on-the-job injuries.

The first step is thorough, effective training for all your supervisors. They need to know "what's in it" for them and their people. They will buy into the project based on this self-interest. Elsewhere in this book I have outlined training classes and programs that cover all the necessary bases, and those outlines should be used in assembling the class or series of classes you will present or have presented to your supervisors. It is very important that upper management participate in these classes and demonstrate that they are a corporate priority.

I believe you should tell the supervisors of your five-year plan and commitment. When they know you are dedicating time, effort, and money to a long-term strategy, they will join you in taking it seriously. The announcement and explanation of a five-year plan eliminates your employees' tendency and temptation to dismiss it as another "fad" or passing interest that, if ignored, will soon go away. Unfortunately, this has been the prevailing attitude of *every* company I have served. The number one complaint is repeated continually:

"Oh, this is a great program, but it will go the way all our safety programs have gone. Trust me, they will preach back safety for a while and we'll never hear about it again. It's like that with every program."

or

"They preach safety but never change anything. In fact, we have traps out here that can hurt us. This program is good but it won't last long because management really doesn't care. If they did, they would fix these dangerous work situations."

What a shame we have a work force that feels that way and management who only give lip service to safety.

By the end of the first year, everyone should have completed "basic training." In other words, all supervisors should have completed both for-supervisors' and for-employees' classes, and all employees should have gone through their classes.

Figure 2. Five-year back attack plan flow chart: year 1 of 5.

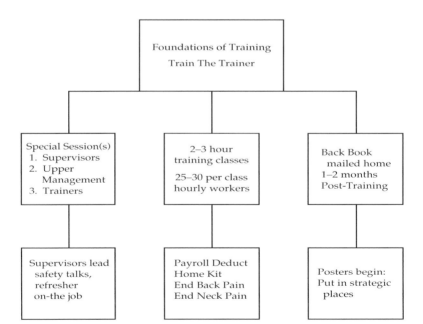

Year Two: Conditioning

Once everyone is on the same wavelength and aware of the same information, you need to work at turning the new awareness into habits of thought, commitment, and action. You can begin successfully implementing corrective programs.

For the conditioning process, I recommend a poster program, frequent safety talks, a flex and stretch preshift exercise program, an awards program, and the formation of an ergonomics committee. Posters should be changed every two weeks or no less frequently than once a month. A good idea is to make posters with photographs of workers at your site, showing incorrect and correct ways to lift, push, pull, and perform various tasks.

Weekly or biweekly safety talks can be delivered in five or ten minutes by supervisors right in the workplace. Here are a few safety talk topics I have found employees receptive to:

- "So You're Thinking Of Starting A Diet…"
- Review Of Risk Factors For Back Trouble
- How Exercise Can Help Or Hurt You
- How To Choose Furniture That Prevents Back Trouble
- Lifting Olympic Style
- Detecting And Correcting Neck Problems
- Buddy Lifting

Flex and stretch preshift and after-meal exercise programs are growing in popularity in enlightened industrial corporations worldwide.[90] Sometimes supervisors conduct the sessions; sometimes employee volunteer leaders handle the duty. Occasionally, a company has a doctor or fitness instructor available to conduct the sessions. Whatever the format, it prevents injuries. People who go in "cold" to do manual labor are at high risk of injury. Simple common sense ought to tell you that, with no expert testimony needed!

An awards program is a very smart addition to your campaign against injuries and for health and fitness. Recognition and reward motivates most people. You may have individual, team, work group, or shift awards. Some award designation ideas include:

- Least Lost Time Back Accidents

- Most Consecutive Days Without A Lost Time Back Accident
- Best Ergonomic/Safety Inspection Of Work Area
- Most Weight Lost
- New Nonsmokers

Let your imagination create as many different ways to recognize and reward behavior that reduces risk of accident or injury and greater work force health and productivity. Allow your work force to help you invent and implement the programs. Nothing is more powerful in effecting change than if they become responsible to and for their own ideas.

The Accident Repeater Program should be used constructively to help each employee understand the cause of his/her accident, but most importantly, the steps they can put in place to prevent future recurrence. This program again should be sold as a benefit to the employees as well as being in the best interest to your entire operation.

The first I heard of an accident repeater program was in 1983 from Mr. Douglas M. Lester, Assistant Director of Health and Safety, Westmoreland Coal Company. If used properly, a program of this magnitude can help you to get injuries under control.[91]

Trainers are ready now for advanced skills and training methods. The superstructure of training is a full three-day program that will empower your trainers and company to stay on course with a five-year back attack plan.

Figure 3. Five-year back attack plan: year 2 of 5.

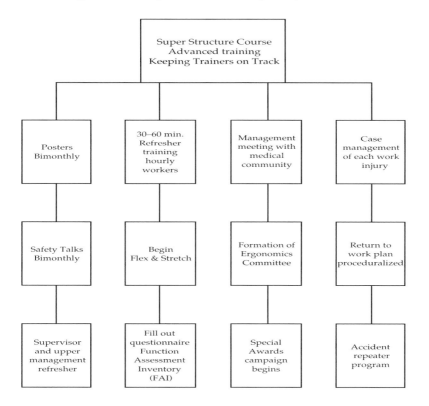

Year Three: Advanced Tactics

In the third year, continue the effective aspects of your attack plan that are already in place from the first two years, and add some relatively advanced or sophisticated tactics, such as a Case

Manager system in which a qualified, competent individual serves as go-between for the injured employee, the company, and the health care providers.

The Case Manager can get to know the employees and preserve the employee's trust and good will for the company. The Case Manager can meet, select, and monitor the doctors and physical therapists who are working to restore the individual's health and get him or her back to work. The Case Manager can set goals for recovery and return-to-work and keep everyone focused on those goals.

Another advanced tactic is to tap into a good preferred provider network. I have set up such networks for the Shell Oil Corporation and the Norfolk Southern Railroad, and welcome opportunities to consult in that area. I already have a nationwide network of physical therapists, for example, fully trained in the American Back School system, so that I can link my corporate clients to providers with the same philosophy and frame of reference.

One more advanced tactic is to conduct one or more meetings each year with your top management, supervisor representatives, employee representatives, case manager, head of your ergonomics committee, and all your health care providers. At these meetings, the "state of the union message" can be delivered, summarizing where the company has been and is, what progress has been made, what problems still persist, and what is being done to solve them. Brainstorming discussions with this group can lead to greater understanding, new ideas, and mutual commitment to goals.

Also, your medical community must be "in synch" with your visions and goals. They must understand your mentality and who you are. Why? If all they have heard about your company has come from injured employees, they most likely do not have a clear perception of who you are. Meetings like this can communicate clearly to your medical community, as well as be fun. We all need breaks from the grind and routine of daily life. If used correctly, this annual meeting can be an oasis.

Figure 4. Five-year back attack plan: year 3 of 5.

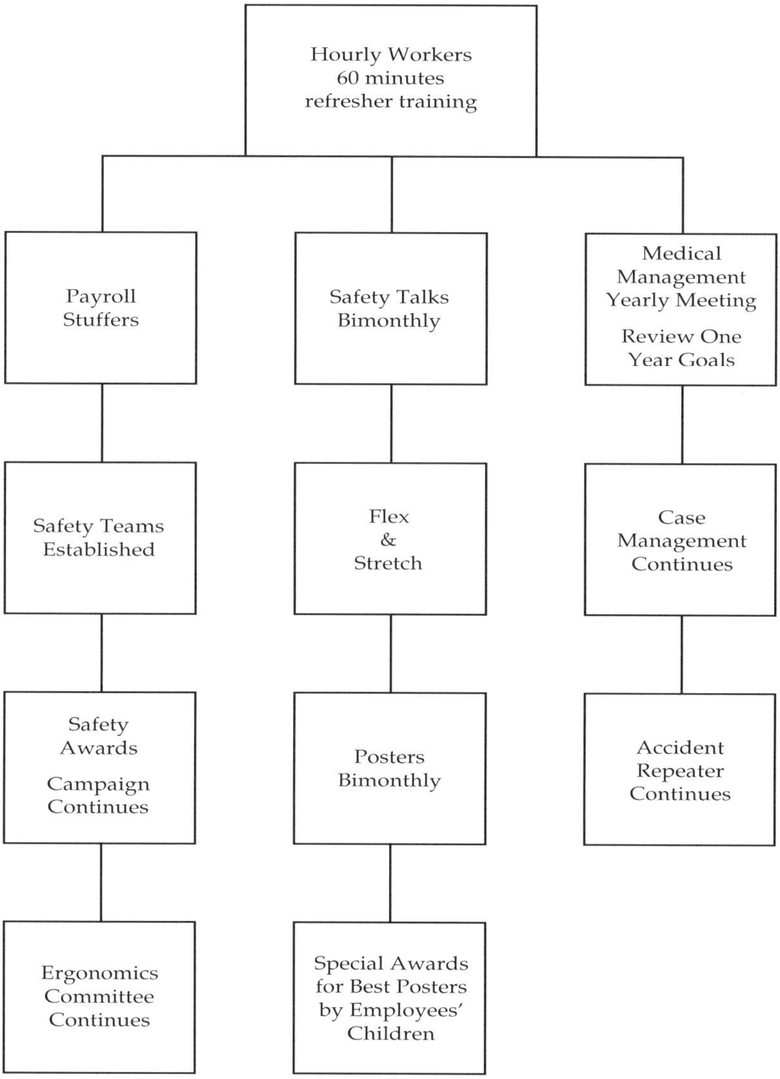

```
                    ┌─────────────────────┐
                    │   Hourly Workers    │
                    │     60 minutes      │
                    │  refresher training │
                    └─────────────────────┘
```

Payroll Stuffers	Safety Talks Bimonthly	Medical Management Yearly Meeting Review One Year Goals
Safety Teams Established	Flex & Stretch	Case Management Continues
Safety Awards Campaign Continues	Posters Bimonthly	Accident Repeater Continues
Ergonomics Committee Continues	Special Awards for Best Posters by Employees' Children	

Year Four: Tactical Adjustments and Major Ergonomic Investments

By the end of the third year, you will have learned and achieved a great deal. You will have seen and measured a trend of reductions in on-the-job back pain, injuries, and lost-time accidents. This will give you the confidence and rationale for making additional, possibly larger-than-ever investments, especially in ergonomic work site changes, to complete the eradication process.

Of course, you will also continue the effective aspects of your campaign already in place.

Figure 5. Five-year back attack plan: year 4 of 5.

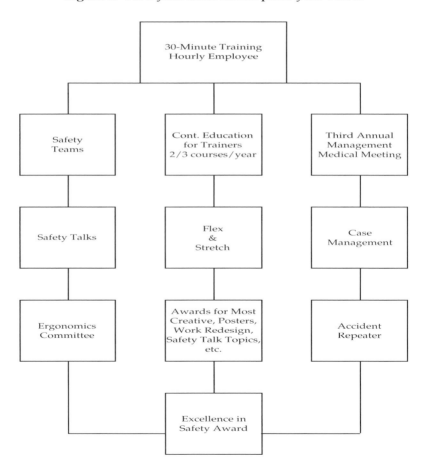

Year Five: Conversion to a "Maintenance Plan"

By the end of the fourth year, you will have succeeded in virtually eliminating back pain as a thief of bottom line profits and worker productivity. You will have a knowledgeable, aware, physically fit work force performing in a safe environment. Now you will want to develop and commit to a scaled-down program of continuing education and information "reminders" to ensure the gains you've achieved remain in place.

Summary of a Five-Year Plan

1. Year 1 — basic training: all supervisors and employees
2. Year 2 — refresher training classes
   - poster program
   - safety talks
   - exercise program — flex and stretch
   - awards program
   - ergonomics committee
3. Year 3 — continue:
   - refresher training classes
   - poster program
   - safety talks
   - exercise program
   - awards program
   - ergonomics committee
   Add:
   - case manager
   - Preferred Provider Network
   - top team meetings
4. Year 4 — Continue:
   - all of the above from Year 3
   Add:
   - major investments in ergonomic job site and/or equipment changes
5. Year 5 — Continue:
   - all of the above from Year 4
   Develop:
   - a "Maintenance Program" for future years

Commitment is the key word for a successful program If you get nothing else from this book, you will profit handsomely just by

understanding and accepting the idea that a thorough commitment of top, middle, and line management to pro-actively attack
problems is necessary and profitable.

Figure 6. Five-year back attack plan: year 5 of 5.

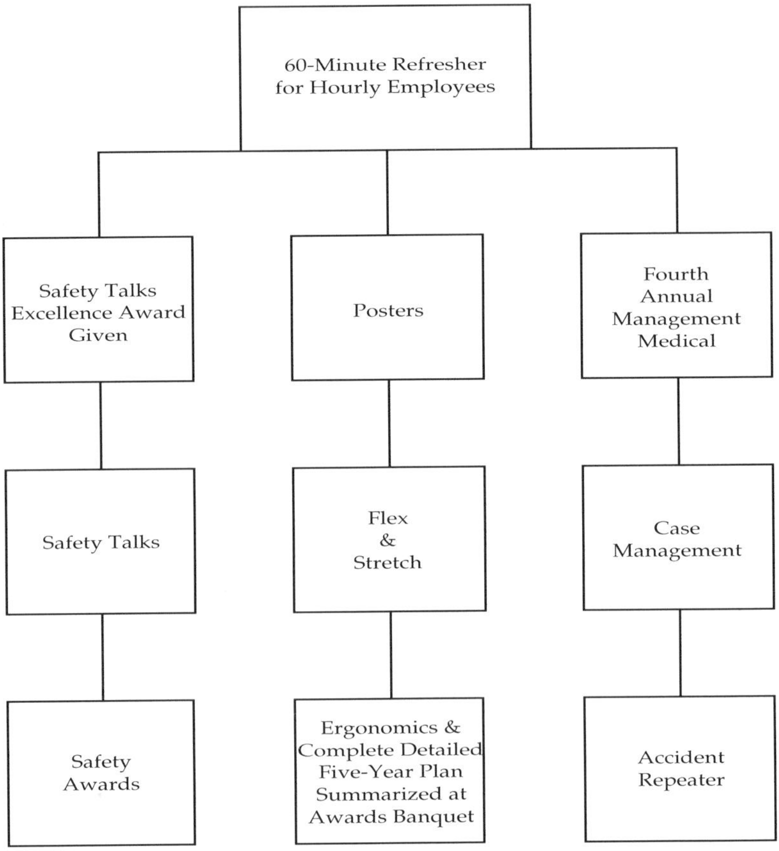

CHAPTER TWELVE

Success is a journey, not a destination.

Ben Swetland

FLEX AND STRETCH

There are really only two things we know as certainties in the prevention of back injuries in the workplace: one is that quality education works, and the other is that physical fitness is essential. The more physically fit your work force, the fewer the injuries.[12,49,69]

Keeping that in mind, go do what *In Search of Excellence* author Tim Peters recommends. He calls it MBWA, Management By Walking Around, and take a look at your workers. What do you see? Lots of "beer bellies" or lousy posture? Sadly, most industrial workers fit this profile: 20 to 40 pounds overweight, unable to do ten good push-ups or jog a half-mile without stopping to rest. Many do no exercise other than the physical work on the job.

A good place to start influencing this situation for the better is with preshift and post-meal flex and stretch exercises. Common sense mandates it. Try to find an athletic team or individual athlete who doesn't "warm up" before playing his sport. He warms up to reduce the risk of injury, as well as to decrease the severity of injury if it does occur. If it's good for the NFL, NBA, American and

National Leagues, and the Olympiads, it ought to be good for your "teams," too.

A side benefit is that some of your workers will take serious note of how much better they feel as a result of this nominal exercise and will seek other opportunities to improve their physical fitness. Super-savvy companies provide exercise facilities and coaches, weight loss, smoking cessation, nutrition, and other fitness programs and services for their workers. If your facility is too small to do this all in-house, arrange preferential discounts and subsidize their workers use of qualified outside providers of these services. The reason I push for a qualified health care professional is that I daily see people who exercise every day of their lives to keep going because they are in back pain every day. Can you imagine that? It's obvious they are either exercising incorrectly or on the wrong exercises. Exercises can feed into low back pain when done incorrectly or without the proper dosage. Again, in 20 years, I've only seen proper exercises cure or help chronic low back pain. So, I always ask these dedicated exercisers if they were taking pills for their back trouble and the pain was still there every day would they continue with the same pills for five years?

It's obvious, once you watch these people, that they feed into their back pain daily and they do not have a clue. We are all so ignorant to our own bodies and our movements that until we are trained to exercise and move correctly we just perpetuate our problems.

The first step is a very safe, simple, ultra-conservative calisthenics routine like school children go through, possibly supplemented with forward and backward arm circles and side bends. Fifteen to thirty repetitions of each movement is more than enough. Done slowly, this will give you a program of 7 to 10 minutes.[10]

It's helpful to have some upbeat music (not rock!) to add fun to the program. Each week, the regimen of exercises should be changed, with the introduction of one or two new, possibly slightly more demanding exercises. If you have no idea how to proceed, go to a teaching supply store in your area and you will find all sorts of helpful materials. Also, photographs illustrating some of the basic exercises can be found on the following pages.

Instantly, many "tough" industrial types are going to discount this as inappropriate for their work force. "I can't see my guys doing a bunch of pansy dance-aerobics," one foreman once told me. This has been implemented successfully at steel mills, coal

companies, and a naval shipyard. And I might point out that some pretty tough football players use ballet as part of their fitness programs!

Another related issue is whether or not your workers are specifically trained and conditioned for the movements their jobs require. Most are not. Most workers are left to themselves to figure out how best to push, pull, lift, sit, etc. in that particular job. If you stop to think, it doesn't make good sense, and the "antidote" is blatant, isn't it?

Different jobs in your workplace require different types of frequent and repetitive movement. Someone skilled in injury prevention and ergonomics needs to study both job sites and workers' actions at each job, so he can prepare instructions and demonstrations of the best methods for safe performance of each function.

The mentality of some corporate management groups is that the loss of even one minute of productivity cannot be tolerated, so giving time to movement analysis, special job performance training, fitness programs, and exercising before shifts and after meal breaks is simply out of the question. To those with such a mentality, I would say your mind is like a parachute: it is of greatest value when opened, not closed! You need to rethink your definitions of "productivity" and "loss" of time versus "investment" of time. We and many industries have found that 10 to 15 minutes of flex and stretch each day increased productivity, and kept people healthier and on the job so they could work and produce. In fact, those companies usually gained a two-fold increase in productivity.

It's very important when implementing a stretch and flex program that you first take the time to educate your employees. Don't just mandate a program and shove it down their throats. People always accept new programs much better when the "why" is explained first. If the "why" makes sense, compliance increases.

And when you stop to think about it, the "why" to warm up your body before work makes a lot of sense.

EXERCISES

In 1960, my little league coach used to tell us, "Boys, stretch what is tight and strengthen what is weak." In 20 years of clinical experience, I have not seen a change in this statement. We all understand that exercise is good, but few of us do anything about it. I believe that the more physically fit and educated your work force is, the less injuries they will sustain.

The following exercises are ones we teach trainers to instruct. As with any program of exercises, you should seek the advice of a qualified health care practitioner before starting. I say qualified because few are versed in how the body moves and compensates for weaknesses or muscle tightness. These simple exercises are generally well-accepted by those practitioners knowledgeable about back care.

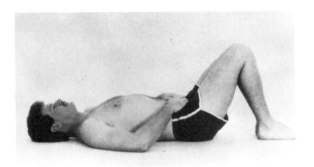

According to Jennifer Locke, R.N., B.S.N., she is using a stretching program to keep industry away from back pain. In her fitness and wellness program for business and industry, she employs the same stretching routines like those an athlete would practice or use before competition. In her article, she further goes on to ask a question, "Should there be any doubt that improving fitness of workers will reduce muscle injuries, especially back injuries?" Once again, we use the common sense approach to the issue of preventing back pain.[90]

This is an exercise designed to stretch the hips, buttocks, and low back region. It should be held for two counts in the finish position for a full stretch, release to starting position and rest for a two-count. Start with ten repetitions, twice daily the first week and increase by five to ten reps a day for the next three to four weeks. Thirty or 40 reps a day is enough. I recommend stretching exercises to be done Monday through Friday, and strengthening on Monday, Wednesday, and Friday. This exercise helps us regain

the flexibility we start losing usually by age 25. Again, the same dosage as for the knees to chest exercise.

When demonstrating knees to chest and prone press-ups, I show the class that ten reps of each exercise will take only 60 seconds. Done daily, they help keep the spine flexible.

Almost everyone has tight hamstrings. This muscle group is usually very inelastic in joggers and runners. Why are they so important? Your hamstrings hook into the bones of your pelvis, run down the length of your leg and insert below your knee. If this muscle group is too tight, every step you take will cause the lower back and pelvic region to create more slack in the system to allow proper walking. When you sit, they will cause you to round your back and, in squat lifting, will now allow you to get into a locked position. If weak, they cannot help in squat lifting. As you can see, they are an important group of muscles. Stretching them is not easy and, as with any exercise, technique is everything. The pictures show a safe way to stretch without involving the back. Use the same number of reps and sets as in the previous two exercises as these are also stretching exercises.

Once a muscle group is stretched, it must be strengthened.

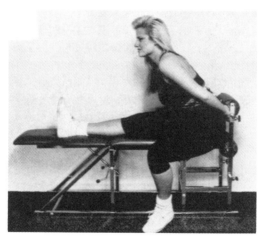

Why? Because of tightness, the muscle has not been able to go through a full arc of motion. If a muscle can not go through a full arc of motion, it cannot possess proper strength. A stretched muscle will be weak if corrective strengthening measures are not taken. To strengthen the hamstrings, you can use ankle weights and curl your leg through a full arc of motion by bending at the knee while standing with the torso and upper thigh straight.

Proper instruction from your practitioner is important because he will keep you from making mistakes in your technique. Almost all of us learned improper techniques in PE, sports, or on our own.

Once proper strength and flexibility are achieved, a lifetime exercise routine should be adopted — a program that meets your needs and takes no more than ten to fifteen minutes two or three times a week. Again, this is not an aerobic conditioning program, but one to keep you flexible and strong.

The next two exercises are for strengthening. Use a Monday, Wednesday, and Friday schedule. Start with what you can handle comfortably. Example:

Week One: 10 reps — rest two minutes, repeat
                  10 reps — rest two minutes, repeat
                  10 reps — stop after this set and forget doing
                             any more until Wednesday.
Week Two: 15 reps — rest two minutes, repeat
                  15 reps — rest two minutes, repeat
                  15 reps — rest two minutes, repeat

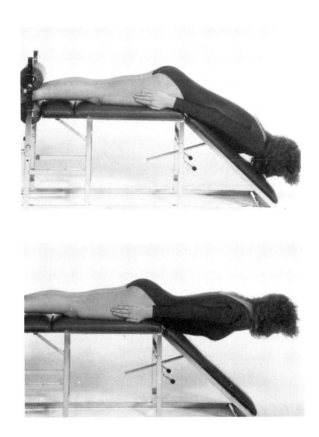

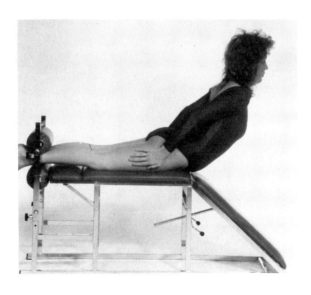

CHAPTER THIRTEEN

Destiny is not a matter of chance, it is a matter of choice.

CONTROLLING MEDICAL COSTS

I am unceasingly amazed at the number of companies exercising little or no control and supervision over the dollars they expend for health care costs. You are signing the checks. You are held accountable for the fairness of treatment in the eyes of your workers. You **must** take a strong position.

It is, in my opinion, an absolute travesty when someone can be allowed to lay off work for six, eight, or even twelve months with no one getting that person into proper medical and rehabilitative care. Yet I see it happening all the time. Somebody is paying for this; why aren't they supervising it? Why should your people have to wait for treatment? It is now accepted everywhere that with early intervention, we have early return to work.

PILLS DO NOT CURE BACK PAIN

Specific to back injuries, it is outrageously common for health care practitioners to keep the injured workers returning to their offices every few weeks, collecting the workers' compensation money while doing nothing but pushing pills or spinal adjustments. This pill-pushing can go on for months. Back pain didn't come from swallowing a pill and it can't be cured with pills or adjustments. Not only is the individual worker/patient at fault for letting this go on, but somewhere there is an employer who has permitted, and paid for, this quackery.[54]

Similarly, if your employees are being seen by a practitioner of adjustments frequently and repetitively for months, you'd better start asking some tough questions. Unlike many physical therapists, I am not 100% antichiropractic. As in most professions, I find there are good ones and bad ones, progressive thinkers and status quo protectors. There are some progressive chiropractors who are aware of the need for rehab, strengthening, and other aspects of treatment beyond the "adjustment," and I applaud them.

CONSUMER BEWARE

Unfortunately, more often than not, I see situations like the 27 year old man who recently came to one of my clinics after three years of frequent chiropractic adjustments, supplemented only by hot packs and ultrasounds the chiropractor erroneously called "physical therapy." We found this young man could not bend forward, and had unbelievably tight hamstrings, no back strength, no abdominal strength, and couldn't even do a sit-up. Because he lacked hip flexor muscle control, he stood crooked or cockeyed when his back pain flared up. After three years and thousands of dollars from his own pocket, he is still in pain and in pitiful condition! Do you think he should have questioned this practitioner sooner? There are millions of people just like him, including your own employees. In just six to eight weeks, with about eight visits to our clinic, and a rigorous strengthening and exercise regimen at home, this young fellow will be back in control of his body, back in shape, back to work, and able to work safely and without pain. His total expenditures will be a few hundred dollars.

If you have employees wandering down a path similar to his, you have an obligation to intervene. You should choose very carefully the practitioners to which you refer and insist they utilize goal-oriented and time-limited programs as well as programs that go beyond pain pills and bed rest or chiropractic adjustments and hot packs. Recently, a study revealed bed rest of two or three days to be sufficient for most back pain sufferers,[56,72] not two weeks! Most people with back pain have been treated as though their case is worse than a person surviving a heart attack!

COMPENSATION CLAIMS MANAGER

I advise companies to have at least one key, qualified individual per 2,000 employees who has responsibility for controlling all workers compensation costs and lost time accidents. Rochester Pittsburgh Coal Company hired a woman to manage cases for their 3,500 employees and she is doing a magnificent job of getting an out-of-control problem back under control. She has forced doctors to go to second and third opinions before they operate. As a case manager, she is monitoring the care of every person who is in rehabilitation and is making sure physical therapists are giving adequate time and attention to truly rehabilitating the individuals. A case manager like her will save your company a fortune.

I also strongly suggest you look to the physical therapy profession for help. Regrettably, physical therapists have stayed quietly in the background and have not widely advertised their existence and capabilities. However, PT's are the best educated musculoskeletal specialists.[63] Physicians, chiropractors, and nurses cannot do what PT's do. They are the only professionals who know how to diagnose musculoskeletal imbalances, screen for risk factors that lead to back pain, and teach individuals how to care for themselves and properly prevent re-injury.

RISK FACTORS

This is only a partial listing of factors believed to contribute to back pain.

1.   overweight — this is controversial and not well-proven
2.   smoking — smokers are two to three times more likely to go to back surgery than nonsmokers

3.   salesman — spend 50% of time driving
4.   frequent lifting — repetitive injury
5.   heavy lifting
6.   frequent bending at waist
7.   weak trunk muscles
8.   weak leg muscles
9.   lack of fitness
10.   inflexible back
11.   back too flexible
12.   previous incident of back pain
13.   leg length discrepancy
14.   disc disease
15.   poor posture
16.   back surgery
17.   tight muscle groups in legs and hips
18.   lack of education on caring for back
19.   psychological
20.   socioeconomic
21.   litigation
22.   trauma
23.   accidents, slips, trips, falls
24.   job reclassification

The list could go on and on. In his landmark article, "Physical Risk Measurements as Risk Indicators for Low Back Trouble Over a One Year Period," Sorensen[69] sheds light on the subject of musculoskeletal dysfunctions leading to backache. When diagnosed, they can be treated and low back trouble avoided.

BACK TROUBLE CAN BE AVOIDED

For the first time in medical literature, we see a complex and credible statistical program that looks at factors within the human body that can cause first time or recurrent back trouble. Backs that are too flexible and poor back muscle strength are at risk. According to Sorensen's article, 15 references are given that sight back trouble as reducing trunk muscle strength.

Both men and women are likely to have recurrent low back trouble if the back and abdominals are weak and hamstrings too tight. Once again, physical therapists can identify those people at risk (as has been done for cardiac disease), intervene with the correct strategy for that individual, and lessen the risk for back pain.

I have written at great length about why I believe physical therapists are the best trained candidates to help you with back injury prevention. If they are one of the best qualified health care practitioners, they certainly should be on your prevention team. However, as a consumer you must make the ultimate choice of just who and what service you will buy.

PHYSICIAN OWNED PHYSICAL THERAPY SERVICES (POPTS)

It saddens me to have to write about this issue, but it is a must. Some physical therapists have believed they could leave a hospital environment and establish their own private practices by working for a doctor. The doctor, under the guise of providing quality physical therapy, will pay the physical therapist a better salary than the hospital and usually provide better equipment. The catch is this: the physician controls the billing process and referrals to the physical therapist. If that's not ludicrous enough, it gets worse. The doctors say they are providing a service for their patients by making sure they receive quality physical therapy.

The fact is, a doctor knows nothing of physical therapy. It's not in their education or clinical experiences. How then would they know quality physical therapy? They would not know quality physical therapy if they fell over it. Their measure of a quality physical therapist is if their patients get well and if the patient likes physical therapy.

Another fact to consider is numerical. If a physical therapist is seeing 20 patients per day at an average of $50 per visit, he earns $1,000 per day, $5,000 per week, $245,000 per year. If the doctor pays the physical therapist $50,000 each year, benefits equal to $11,000, $20,000 ancillary help, $3,000 per month for leased equipment and $1,500 per month for rent, that comes to a total of:

$50,000. PT salary
+$11,000. Benefits
+$20,000. Ancillary personnel
+$36,000. Equipment lease
+$18,000. Rent
$135,000.

Now subtract 90% of total billing (some accounts you will never collect):

$220,500.
–$135,000.
$85,500. Clear profit

This has been a very conservative estimate, too! Imagine no equipment lease in five years or no building lease in seven years, add another $54,000 to the bottom line.

Why this scenario? To point out that doctors can profit from $85,000 to $100,000 per year per therapist they employ by referring their patients for "quality physical therapy." What doctor could resist such easy money? Could you? You tell your patient to go down the hall or next door to see your physical therapist and that's all the work you have to do for a cool $100,000.

As you see, there is a critical question of ethics here. The American Physical Therapy Association has issued a stand and they are against all referral-for-profit situations. Representative "Pete" Fortney Stark (CA) has introduced the Stark Amendment to abolish such unethical, immoral practice. He writes of such atrocities as sending very sick and infirmed patients to an imaging center 50 miles out of the patient's way because of part ownership of the referring doctor. Imagine sending a sick person 50 miles out of his way just so you can make an extra buck!

I object to all referral-for-profit situations. The ethical question is who serves whom? Who has taken an oath? It seems that in this referral-for-profit situation, patients' rights are stripped away for the sake of the almighty dollar and ethics can be cast into the sea.

We have tried to educate those in our profession about Physician Owned Physical Therapy Services (POPTS). It is still a raging controversy that must be legislatively determined to be unlawful.

The issue is is how much the physical therapy source will pay the doctor(s). In my area, there is a group of nine orthopedic surgeons from a neighboring state who opened an office. Each doctor spends four hours a week in the Ashland, Kentucky office. Obviously, I desired their business as they had no physical therapy department in the Ashland office. However, they did have a PT department in their Huntington office. Once they saw the service I was rendering, they referred quite a few patients. However, in September 1986, they decided to open a physical therapy office in

Ashland, as I was too far from them (all of a three-minute drive). Guess how many patients I've seen since then? That's right, very few. In all fairness, there is one of the nine doctors who has sent maybe six or ten people in over four years. Now that's a stark contrast to 35 referrals/week prior to opening their physical therapy clinic. I do not know why physical therapists such as those who serve the orthopedic surgeons can be so ignorant, naive or just plain stupid.

I have three friends with similar stories. One in Louisiana had a group of eight orthopedists sit down with him and his four partners (eight MD's vs. four PT's) and say they were not profiting from their referrals. They did not care about service or quality, all they wanted was 40% of every dollar my friend collected on patients referred by this ortho group. With all due respect, my friend told this group to forget it. The ortho group then hired two physical therapists and no longer sends my friend patients after 15 years of exemplary service.

Another friend lives in the Washington, D.C. area. He sits on Insurance Boards as a reviewer of physical therapy services and helps the insurance company decide who to pay and who not to pay. He tells me 96% of all orthopods in that area own their own physical therapist. He also tells me of denying claims for as much as two or three years of hot packs, massage, and ultrasound for low back pain patients.

Imagine seeing a patient for two or three years, three times a week. The patient should be a better consumer than that! Certainly the employer or insurance company paying such a ridiculous bill is. Guess what happens when these claims are denied? The quality physical therapy service is fired! This is happening right in our nation's capital! I ask you, are these MD's above the law?

Yet another friend in California tells me of one of his physical therapy colleague who was in a POPTS for ten years. He made $85,000 to $100,000/year working for his best friend, who was an orthopedic surgeon. They were educated together, their families were best friends and they kept this work situation for over ten years. But the surgeon divorced, losing half his belongings. He fired is so-called best friend, hired a young kid out of physical therapy school for $35,000, and pocketed a cool $65,000. So much for quality physical therapy and friendship.

The picture I've painted is one of unattended abuse. It's like throwing a thick, juicy steak to a hungry dog and telling the dog not to eat. It's a rare dog that won't eat. It's a rare situation of referral for profit that won't be abused. As a consumer of health care, you must buy your services wisely. Ask your physical therapist if he is financially affiliated with doctors or if the doctor is financially affiliated to "his" physical therapists. It's your right. If they are, you are most likely paying for abuse.

CHAPTER FOURTEEN

The speed of the leader determines the rate of the pack.

Keeping Trainers and Supervisors Motivated

One-shot training is not a very good idea. The old saying "out of sight, out of mind" applies perfectly to back injury prevention in the workplace. If we stop paying attention to it, people will lapse into old, bad habits. Laziness will take precedence over safety. Work sites will deteriorate. New, untrained workers will infiltrate the workplace and the injury problem will be given new life. Trainers and supervisors originally involved in your program to stop back injuries at work must be motivated continually to "keep the message alive."

My biggest recommendation is that your team of trainers and supervisors be committed to annual, quarterly, maybe even monthly goals, and be given significant recognition for accomplishment of these goals and contributions to the corporation. I find that trainers often suffer from a lack of recognition from management and respect from peers and workers. Sometimes the hourly employees look at them as having "cushy" jobs of little importance.

TRAINING IS A TOUGH JOB

Training, incidentally, is a tough job, as anyone who has ever stood in front of a group for eight hours and poured himself into his material will tell you. Zig Ziglar, one of the most prominent, successful motivational speakers in America today, says after he has spoken for three hours, he has burned up the same calories as are burned in eight hours of manual labor. For this reason, as well as for their value as an example, trainers should be physically fit and they should keep themselves fit. They should also be given a few days away from the job at least every few months, to re-energize and rest. Having them attend refresher or new courses, or trainers' conferences is a good way to give them a break and have them return with renewed enthusiasm for their jobs.

The National Safety Council, American Society of Safety Engineers, or the American Society of Training and Development all offer a number of conferences and seminars. As a benefit for the trainers and supervisors, I recommend you pay their membership in these organizations. Any of these forums will provide many opportunities for your trainers and supervisors to network with their peers in other industries, share ideas, and develop new ideas.

BRING TRAINING TO YOU

If sending your trainers and supervisors to distant conferences is beyond your budget, you might look for local opportunities for them to gain motivation, experience and contacts, even if it's the famous Dale Carnegie Course or a Toastmaster's Club. You should also provide them with a steady flow of educational and motivational books and audio cassettes. All sorts of good books and tapes on public speaking, training, motivation, and management are readily available in local bookstores as well as from companies like Nightingale-Conant Corporation (Chicago), Empire Communications Corporation, publisher of this book (Phoenix), and my offices.

You can also bring seminar leaders to you. We call it "corporate buyout." We require 25 to 30 people so there are times when one sponsoring company will network with other local companies to fill a class and share the cost. You can give your trainers and supervisors an opportunity to collaborate creatively, gain respect and recognition for their contributions, and continually educate and motivate the troops by having them put out a monthly or quarterly "safety newsletter."

Lastly, your trainers must be part of your company's "big picture." You may find value in including them in certain corporate meetings, so they fully understand the importance of what they do. If back injuries are 80% of your compensable lost-time accidents, for example, they may represent millions of dollars. They need to understand this in relationship to other budgetary items.

If you want them to get super results for your company, you must give them appropriate support. As shown earlier, Lockheed[71] found they saved $6.50 for every dollar invested in training — That's 650%! Training pays. Take good care of your trainers — they are worth their weight in gold to your company.

Supporting your trainers means showing or leading them in how to find time to embark on yet another job function. Believe me, the number one complaint of all trainers is, "How will I ever find time to do this program?" By mandating and giving them time in their schedules or acquiring help for them, you will unburden their minds, enabling them to concentrate on the safety project and reap your desired results. If you expect your trainers to do this and all their other duties, I'm afraid your program will suffer and your results will reflect it. Again, my job, in writing this, is to ask you to expand your thinking and work creatively to solve your problems.

CHAPTER FIFTEEN

All things are difficult before they are easy.

John Norley

CHANGE IS CONSTANT

One lesson we all should have learned by now is that change is a constant. No two days are ever alike. Tom Peters, now an international figure in business since publishing *In Search of Excellence*, has a sequel entitled *Thriving on Chaos*. Frankly, the second book is so thick I've postponed reading it. However, *In Search of Excellence* is nothing new! He who manages best, wins! Big surprise!

Just the other day, I received a phone call from a man who wanted to schedule an appointment to see me because he hurt his back at work the day before. As we talked, I thought of a few questions and asked him:

"Have your reported this injury?"
"No."
"Have you been to the hospital?"
"No."
"Have you been to the doctor?"
"No."
"Why?" I asked in amazement.

His answer floored me. He said he did not want to report the injury because his supervisors were a bunch of jerks and the powers to be would give him a hard time, if not fire him. The mentality of management and employee in adversarial roles amazes me! How counterproductive! I informed him that, for any medical doctor or physical therapist to treat him for a worker's compensation case, he would have to file. Otherwise, he would pay out of pocket.

ATTITUDINAL CHANGE

For those who maintain a hard-line stance against back pain or those looking for positive change and cost containment, please read Fitzler and Berger's report, "Attitudinal Change: The Chelsea Back Program" in the February 1982 edition of the *Journal of Occupational Health and Safety,* page 24-26. The authors describe how they define reorganization change

1. structurally
2. technologically
3. attitudinally

As the title indicates, a management attitudinal change took place, case management implemented, immediate care given to work injuries and, in a period of five years, Eureka! — almost total eradication of back injuries. Go ahead and read it.

That article was written eight years ago and we still see mismanagement of back injuries from beginning to end. If you have stayed with me up to this point in the book, I know that change within you has occurred.

SOCIALIZED MEDICINE IS NOT THE ANSWER

We must understand that we control our future. As an entrepreneurial nation, it should frighten us to think of more government intervention in health care. Some scream out of total ignorance for socialized medicine when all of the world looks to America for hope and help. Socialized medicine does not work! If your local doctors don't believe that, send them to Great Britain. I doubt they would go. If they did, they would not stay.

We need to exert case managed control over each injury and work concertedly with all involved health care practitioners for a time-limited, goal-oriented, return-to-work program. I contend that if we let this slip and offer it to the U.S. government, we will pay billions for truly mismanaged care.

Are you convinced yet? Are you convinced that back pain does not have to be a cost of doing business? I hope so. As prevention managers, we must form a network among all industries willing to strive for change and cost containment. We must network the health care practitioners who wish to be a part of the solution. We must network with elected officials for a change of antiquated acts,[64] such as Federal Employers Liability Act (FELA, 1906), under which railroads fall, and Federal Employers Compensation Act (FELA, 1916) insuring the government civilian employees and state worker compensation acts. Why? It's simple. We can not and should not bear the burden of antiquated legislation in our litigous society.

ABUSE ABOUNDS — YOU KNOW IT AND I KNOW IT

Why should a government civilian employee be guaranteed 45 days of continued normal pay after injury? This just begs for abuse. I wonder how many miraculous recoveries occur on day 46 post injury? And you and I wonder why taxes always go up? What about states that pay an injured employee $350 to $450 tax-free compensation each week after a work injury? Obviously, some people have no financial need to go back to work. The system does work for those who need it — they should receive worker's compensation benefits. But abusers of worker's compensation spoil the situation because they are allowed to.

With such flagrant abuse of the system, we, as a society, must rethink what we are about. It is unfortunate that we have generations within a family that have never seen a mother or father go to work. How can future generations ever understand work ethics?

We must always take care of those who cannot care for themselves. But we must never reward childish behavior with a free ticket under worker's compensation, FELA, or FECA. As Zig

Ziglar would say, "There is no free lunch." We must vote and vote again for those who think and believe as we do. We must also watch them once they are in office. If they do not stand for the reforms for which we voted them into office, we must ensure they never get re-elected.

As a little boy, I listened to a young upstart who was elected president of the United States of America. In that bitter cold January of 1961, President John F. Kennedy, in his inaugural address, called for all citizens to "ask not what your country can do for you, ask what you can do for your country." We have it too easy here in America. Most of us eat like kings and queens every day. We are struggling about what to wear because the closet is so full while three fourths of the world's population could never imagine all we have. We have become too complacent! We do not even vote, our unalienable right as U.S. citizens! Millions have fought and died so we might taste the democracy and freedom we enjoy. I only hope they are not turning in their graves.

PRE-EMPLOYMENT SCREENING

Many variables are under our direct control. Historically, pre-employment screening has been a low back X-ray, cursory physical, and then off to work. The physical, while nice, only told us about heart rate, respiration, blood pressure, urinalysis, and blood work. X-ray's of the low back serve no purpose in pre-employment screening. Andersson, et al.,[4] concludes that low back X-ray's have low sensitivity, low specificity, and low predictive value for low back trouble. Then what should we expect from a pre-employment selection program? According to Andersson, et al., there are four pertinent questions to ask:

1. Is the screening safe?
2. Does the screening procedure have a good probability of predicting the risk of low back pain or injury?
3. Is the screening procedure practical?
4. Is the screening procedure ethical and legal?

He poses other questions for predictive value:

1. Is the test scientific?
2. Is the test specific?
3. What is the frequency of positive results in the population?

Of course, another question to answer is whether the screening program is practical and cost effective.[4]

Strength testing, while controversial, should be a minimal part of your total program. The applicant should have strength at least equal to the tasks required of the job he seeks. We must place workers in jobs that they can handle while not placing them at risk. Job redesigning can be expensive, although it can increase bottom line productivity of man and machine. Controlling what is to be lifted, how much, how heavy, and job site design all appear to be reasonable and necessary efforts for controlling occupational low back pain.[9,68] As previously stated throughout this book, training is a must for those in manual handling situations.

FUNCTIONAL ASSESSMENT INVENTORY

There exists an instrument for use on existing employees that can help you zero in on their perception of the cause of low back pain in your industry. It is called the Functional Assessment Inventory, designed at Washington University in St. Louis.[60] This instrument was used with coal miners, enabling us to determine job tasks that put workers at risk, offer technical changes, and gather many other interesting pieces of data, such as make-up of the work force with regard to age, sex, weight, height, number of years in coal mining, etc.[59]

We also found, in two separate surveys, that 88 to 92% of underground coal miners work with back pain daily. Obviously, this is disproportionately high! We surveyed over 2,500 coal miners and reported on a selected random sample of 799. Now the coal industry has information to use to effect outcome and cost containment, and decrease human suffering.

WARM UP AND DECREASE SPRAINS & STRAINS

As previously mentioned, exercise programs have been found to be effective. It will cost you five or seven minutes before each shift and after a meal break. The expense is nothing compared to musculoskeletal accidents. Immediate implementation is highly recommended. Exercises are general calisthenics like those learned in public school. They should be simple and safe, effectively speeding blood circulation, increasing heart rate, and stretching muscles and ligaments.

TOTAL LIFE CONCEPT

Some companies have gone on the kick of a wellness campaign or a health promotion program. According to Spilman, et al.,[70] AT&T conducted a pilot study of a health promotion program and spent considerable time considering their needs. Through using this strategy, they applied tactics to bring about change. The Total Life Concept (TLC) was put together to create a work environment supportive of positive health practices. They did not just talk about it, they did something about it. They can statistically show changes in their employees' dietary and smoking habits as well as improved perceptions of their own health. People must assume responsibility in their own care. Another wonderful benefit was that the employees who participated in AT&T tactics had a significant increase in their belief that AT&T was interested in their welfare.

After reviewing the article, it pushed a button in my memory. Mr. Freddie Howe, mine superintendant at Wolfcreek Collieries, told me, in 1984 after training his people, "If this program doesn't do one more thing for us, it has worked. Since training, our morale has been better than it has been anytime during the past two years."

I have always presented this program as a benefit the company provides to its employees. The company did not have to hire me! I think everyone is surprised at how much they learn about caring for themselves while having fun at the same time. With Spilman's article, corporations like Johnson & Johnson, Tenneco, and others, we will see benefits reaped by efforts put in. A law of life is that you can only reap what you sow. As my dad used to always tell me, effort in equals effort out.

PREFERRED PROVIDER NETWORK — PPN

Our Preferred Provider Network links you to health care practitioners with an expertise in industrial/occupational musculoskeletal dysfunction. They have a desire to work with you to get your problems under control at a fair and reasonable rate. They have been through our health care practitioners' seminars and use the ABS equipment daily in their clinics. They also have direct access to me.

This program has already been proposed to Pepsi, is in place with a division of Shell Oil, and is getting started at Norfolk Southern Railroad.

OBJECTIVITY IS A MUST

The days are over when a doctor, chiropractor, or physical therapist can keep an employee off work simply because of doctor/ therapist status. Without supporting objective and subjective measures to back their statements, a health care practitioner is on thin ice. You should never just take his word for it. If the request of one or two weeks bed rest is made, it should be shown from the literature how it will benefit your employee. In fact, it is contrary to Waddell's finding that two to three days bed rest is sufficient. Ask for the articles that support diagnosis and treatment. You can read it yourself and, if it needs interpretation, that is what you are paying the physician or therapist for. We now have computers that reliably and objectively measure strength, endurance, and total work done by various muscle groups in the arms, legs, back, or abdomen. Yet there are other tests that can be done without fancy equipment that give you objective quantification of one's ability to work.

WORK HARDENING (WH)

Perhaps that is why work hardening programs have caught on like wildfire.

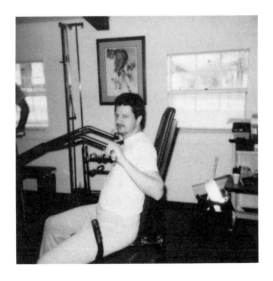

Consumer beware. I have heard of rip-offs that are a nightmare and give work hardening a bad name. There have been reports of patients lying in a practitioner's office all day, supposedly on a work hardening (WH) program. The bill could be as high as $650 to $700 per day — blatant fraud and abuse. In fact, it was so bad in the state of West Virginia that a one year moratorium was put on all WH programs. It is all across the United States. Some believe WH to be exercise programs or exercise circuit training on fancy equipment like nautilus. It is not!

Once you become involved with a true WH process, you will see its benefits. With objective documentation and goal-oriented, time-limited programs, you should be able to expect, measure, and predict outcomes.

SPORTS MEDICINE

I believe work hardening is a misnomer. Work readiness or work preparedness is a more appropriate name. Years ago, we found many patients were not helped through traditional means. We basically took from sports medicine a program of consistent, safe loading of the body to flexibility, strengthening exercises, circuit training, tread mill, bike ergometer, stair climbing, etc. The injured employee is taught how to pace himself. He is responsible in setting goals, achieving the goals and recording everything he does. We may not always achieve 100% pain cessation; however, we do teach them (as Klein and Sobel state in their book) that they can and should be proactive in their care.

The focus of a Work Hardening program should always be an improved function, not pain reduction. It is typical, however, for us to see pain reduction with improved function. There should also exist a clear, open line of communication between your case manager and the work hardening coordinator. I recommend at least a weekly review of progress or lack thereof.

NEW MODEL

Dr. Tom G. Mayer, et al., has reported to us of a new model for empowering chronic back pain sufferers and post-spinal surgical patients returning to work.[86]

This program is unlike pain clinics in that the participants were involved in a rigorous, three-week rehabilitation program for 57 hours a week. They promoted functional restoration through specific exercises, training and functional tasks, education, and work simulation/hardening. They also provided psychological intervention as well. This program is controlled through four phases so that upon discharge they would go through controlled follow-up and monitoring.

At the end of one year, 85% of the treatment group were either working or in a work training program. This is in contrast with the nontreatment group where only 39% were working. From this study, we see that there is a four times greater unemployment rate for the nontreatment chronic back pain group. They even did a two year follow-up and found the same trends as in year one.

Another model of a comprehensive program exists in Dallas, Texas, at P.R.I.D.E. An explanation can be found in the article by Peter B. Polatin, M.D.[87] He proposes a wonderful alternative to traditional therapies for those with chronic back pain. His measure of success is for patients to return to work and remain at work over one year. Two-year follow-up has been 80 to 85%. Although "chronic" is not defined in the article, he does give two references for the statistic.

Perhaps you have already experienced work hardening. If not, visit one near you. Ask tough questions and see how the program may fit your needs.

According to one of the world's leading authorities, Dr. Alf Nachemson, who believes in work for all, even those with low back pain. He states in his article that with our new information that we have on diagnosis and treatment for back pain early, gradual, biomechanically controlled return to activity and work should be in place upwards to 80% of patients with low back pain in whom no objective causes for pain can be found. Furthermore, he states that patients must be told repeatedly that a gradual return to work will not worsen their condition in the future. However, they must be warned of recurrent back trouble.[89]

AN EXAMPLE OF BEING RIPPED OFF

During November, 1990, I trained several trainers from various coal mines in Elko, Nevada. While training at Barrick Gold Strikes facility, I was asked to look at a young man who had just returned from Salt Lake City where he had been in a work hardening program. During the two or three weeks of the program, he was housed, fed, and trained to recover from his work injury, at the cost of $12,000.

I asked him what he had to do and he showed me. I then asked him how he should lift, push, pull, and sit. He bent over at the waist to pick up core samples in a box to be transferred to shelves for storage. The box weighed 35 to 40 lbs. He should have lifted it with the back locked in and pushed the load up with his hips and legs. His hamstrings were so tight he could not bend over without bending is knees. He had weak legs (he could not squat) and a hypermobile back, which also was probably weak.

As ludicrous as this sounds, it gets worse. He told his work hardening doctor he enjoyed mountain biking. The doctor told him it was great for him because it would stretch his tight hamstrings and strengthen his back! Now that's as bad as it can get!

Let's look at it. In riding his mountain bike, he is bent over or slumped in his spine, causing his back to be rounded and put the muscles on stretch. This causes muscles to weaken. Furthermore, his back does not move. To strengthen back muscles, you must move the back through its full arc of motion with resistance, just like you would strengthen the biceps by doing curls. Yet, you cannot do that on a bike. Also, his hamstrings could not possibly be stretched because his knee is always bent to pedal the bike.

The young man still hurts, does not understand why, does not know how to take care of himself and the company thinks he has a bad attitude because he still complains of pain. After all, the company has done all they know to do, so they think he is a bad employee. I've got news for you — the company was robbed of $12,000 because they are ignorant of what they should be getting. Then they absolve themselves of this employee problem because he was sent to work hardening.

My recommendation was obvious: never use the facility in Salt Lake City again. After training, the trainers could teach the young man proper movement patterns for bending, reaching, lifting, pushing, pulling, and sitting with a low back support. Another suggestion was that the company should make team site visits to various work hardening programs and see first hand what their people would be getting. Lastly, the young man should have known how his injury occurred, and how he could have prevented it. He should have no musculoskeletal defects, or at least be on the proper stretching and strengthening program. By rote, he should have been able to do all his activities correctly.

If this were an isolated incident, I would not have written about it. The problem abounds and you must beware by becoming aware. Otherwise, you may presume you have done everything for your employees, too. Unfortunately, too many employees have been written off because some practitioner could not find the problem on X-ray, CAT scan, or some other test.

No, work hardening programs may not get rid of one's pain. But they should enable a person to cope and do some kind of meaningful work. Pain is not a reliable measuring tool for success of a program — at least not in the realm of back problems. Work hardening programs should bridge the gap in traditional medical approaches for the injured industrial worker.

STATE INTERVENTION

Finally, we may see state intervention. In Kentucky, worker compensation regulations have been revised to control LTAs and facilities wishing to participate in treating work hardening patients must have state or CARF approval.

Oregon's legislature has made changes in the workers' compensation system that go beyond reform and create a new system for insuring workers who suffer job related injuries. They are sweeping changes and reveal a shift in philosophy from one of encouraging workers to stay home and collect benefits to one of workers returning to a productive life. The new system will increase benefits for severely injured workers, control medical costs, reduce litigation costs, improve on the job safety, eliminate awards for injuries not truly work related, increase job reinstatement rights for workers, and provide incentive to employers who hire recovered workers. They estimate a significant reduction in the $800 million that employees paid in workers' compensation premiums in 1989. It is much too early to see what these savings will be as this has just been put in place. As I said in a previous chapter, there has to be control of fraud and abuse in all parts of this system to better control low back pain and other work-related accidents.

STOP ABUSE — YOU CAN DO IT

All Oregon has done is eliminate fraud and abuse. It tears me up to see they will not pay benefits to injured workers who are in jail. Benefits will also be denied to workers whose injuries are proven to be the result of alcohol or illegal drug abuse, unless the employer encourages or has knowledge of such abuse. Sounds strange, doesn't it? Hopefully there is an explanation for this.

In summary, many changes will have to take place in order to truly curb this epidemic of back pain. As trainers, we can have a most significant impact on employees and prevent them from ever getting into the quagmire of workers compensation, medicine, insurance, disability, illness, behavior problems, and other destructive problems. Let us work on those things we can change.

SAFETY TRAINING

Now let's turn to safety. Touchy subject? Training occurs once each year? Once in the lifetime of any employee?

I don't care how good you are as a trainer, a one-time dose of back education will net a one-time dose of back education. Effort in equals effort out. What would happen to McDonald's if they advertised once a year? Could Burger King ever make it without the constant reminder that you can have it your way? Something as simple as being flexible with hamburger toppings has created a monster industry called Burger King.

Just dream with me for a minute. We know the human mind needs constant bombardment in order to understand and act on information, hence advertising agencies. What if your upper management team were to create a learning atmosphere that was constant, daily, ongoing, creative, and fun for controlling lost-time accidents? We are talking radio, TV, bulletin boards, newspapers, and newsletters that bombard their employees with information that will encourage and enable safe work habits.

YOU MUST PLAN

Whatever you do, you must plan to control your own problems. By using the strategies in this book, I believe you can do it. By choosing your tactics and implementing your strategies through a five-year plan, you can come out on top.

SUMMARY

Throughout this book, I have tried to present facts, ideas, and methodologies. Not all studies are correct in their hypothesis, testing, predictions, or outcomes. However, I have presented both current and old literature.

Some of the old myths of back pain have been shattered. Bed rest used to be the standard treatment for back pain. According to Waddell and the Quebec Study,[73] two or three days bed rest is recommended for acute back pain. Imagine how many billions of dollars of productivity have been lost due to the nonsense of putting people to bed. Back pain sufferers have been treated as though their cases were worse than surviving a heart attack.

I challenge you to challenge the future, to challenge authority when it makes no common sense (like two weeks bed rest for back pain). Work as a team, form a plan, and work your plan.

EPILOGUE

As with any program, you must market it. You must have a campaign for the program and commitment from top management to hourly workers.

How many millions, if not billions, of dollars have soft drink manufacturers spent on advertising just to get you to buy their product when you are walking down the store aisle?

Answer the following:

- Coke is the _____ thing.
- _____ is the choice of a new generation.
- _____ is the uncola.

If you have been looking for the real thing, Coke would have you to believe it can be found in a can or bottle with their label on it. How many times have you heard Coke, Pepsi, and 7-Up commercials over the past 20 years? They are on TV, radio, magazines and billboards. What about these slogans?

- You deserve a break today, so get up and get away to _____.
- Have it your way at _____ .
- _____ we are out to win you over.

How could you possibly deserve a break at McDonald's, have it your way at Burger King, or have Hardee's win you over without mega-advertising? Get your employees involved in your back-attack plan. Have them become a part of it — or even begin it — and they will see it through to success.

APPENDIX

FIGURE A.   The possibilities are up to your creativity. I would suggest expensing out money for: 1. A palletizer. This allows the worker to keep this pallet at waist level even when the load to be lifted is on the bottom of the pallet. 2. You could keep a fork lift available and it could be used as a palletizer. 3. Make sure the workers lift with their backs locked in and have excellent leg strength. 4. Jib hoist. 5. Smaller containers.

FIGURE B.   Keeping the trainers classes small has been a key in getting excellent results. Here the author is instructing a group of trainers. With back injuries being such a serious issue it deserves our best attention. With a large class the intensive interaction is hampered. Your trainers should likewise keep 20-25 employees per class.

FIGURE C.   Here on the afternoon of Day 2, a trainer delivers a part of the educational format. We have found this helps them hurdle the barriers of presenting this program quickly. It allows for constructive positive critique by the author and other classmates.

FIGURE D.   A group of advanced graduates after a tour of Physical Therapy Center at Ashland.

FIGURE E.   Advanced trainers going through musculoskeletal risk screening. Here they are testing the length of hamstring muscles.

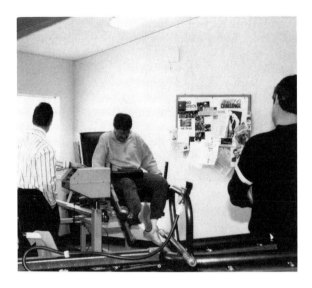

FIGURE F.   Advanced trainers testing their strength on computerized piece of rehab equipment for testing the strength of the knee, as seen here, but it can also test hip, shoulder, ankle, and elbow joint motion strength and total work done.

FIGURE G.   Testing the abdominal and back strength on a isokinetic computer. This is like the machine in F but it tests only the trunk muscles. This machine also tells us the coefficient of variation of one repetition to another, and we can discern who is trying to control the test and who is being cooperative. With these new computerized devices we can use objective reliable date to help screen, test, and rehabilitate.

FIGURE H.   The author is training a small group of steel mill workers near his clinic. Small groups of 20 have worked well. This steel mill has contracted for two-hour classes and on the job follow-up. Here the author explains the components of the spine and how they work in laymen language.

FIGURE I.   The one thing the author has learned is that the workers need to participate by example. During a two to three hour class the author has the participants up every 10 to 12 minutes, exercising, bending, lifting, pushing, and pulling.

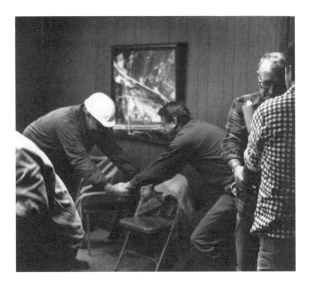

FIGURE J.   Here is the pulling example. The employees always love this example because they can immediately feel the difference between back locked in vs. back rounded out. It all has to do with using what you have and leverage. This is the most popular of all demonstrations.

FIGURE K.   Training can be ! If we make it that way.

FIGURE L.   The author as a keynote speaker kicking off a health and safety campaign for a large petrochemical company.

FIGURE M.   Possible solutions from pages 81: 1. Build the box two to three boards high. 2. Put box at waist level and sturdy stand. 3. Put hinges on front panel. 4. If nothing else, have employee attempt back locked in position, however, in this posture it is impossible to do, at best he may use some back muscle. 5. Flex and stretch prior to working this job.

FIGURE N.   Possible solutions: 1. Jib hoist. 2. Put barrel on tilt stand waist level. 3. Stretch and flex prior to doing this job.

FIGURE O.   Possible answers: 1. The heaviest objects are above shoulder level. 2. All objects to be lifted are well over 120 lbs. 3. No mechanical assist device.

FIGURE P.   Possible answers (from page 85): 1. Always have two people lift each piece of steel. 2. Use a small jib hoist (the cost is a fraction of one back injury). 3. Always use "back locked in" technique for lifting. 4. Flex and stretch prior to work and after lunch. 5. Keep metal off the floor — use more benches up against wall when fan is wide open.

FIGURE Q.   (from page 85): Until ergonomics kicked in at this facility, all motors in the roll mill were set in deep as you can see in the middle of this picture. The employee had to lift this motor over quite a distance and reach out with arms extended over the platform barrier. This motor weighs 150 lbs. Is it any wonder people were hurt while replacing motors?

FIGURE R.   (from page 56): The solution was simple and it was accomplished by resetting the motors out to the platform by using longer metal shafts. Will this save back injuries? You bet! Why do we always wait until someone is badly hurt before we look to change potentially dangerous situations?

FIGURE S.   Possible answers (from page 86): 1. Bend hips and knees. 2. Lock back in. 3. Push load up with hips and legs. 4. Put bin on angle at his level (3-4″ below waist).

FIGURE T.   Answers (from page 87): 1. No, he is not protecting his back. 2. Two person lift. 3. Use locked in style of lifting. 4. Stack bolts in smaller bundles, less stress/lift. 5. Use mechanical equipment such as scoop and slide from supply car to scoop and avoid lifting.

FIGURE U.   (from page 87): This is a million dollar piece of equipment. It has been given 25¢ worth of seat and operating deck design. Only the consumer of this product can force redesign.

FIGURE V.   Our female subjects lifted 15 to 30 lbs. The author recommends lifting with one foot in front of the other in this style of lifting so you can straddle the object and keep your feel flat for a stable base of support. You should never lift on tip toes as this creates too much pressure in the knee. If you cannot keep flat footed, the heel cords and hamstrings must be stretched.

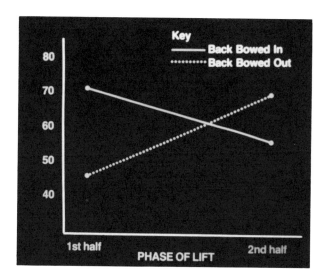

FIGURE W. Graph showing how back muscles work with back rounded out vs. locked in. The locked in method was also found to be the preferred style for lifting amongst test subjects. Once learned it feels comfortable and strong.

FIGURE X.   This Thailand lady is working at a loom daily with back rounded out. There are those that still believe this is the most desirable position to lift, work, and reach, as they feel it is a stronger position.

These are just a few examples of the use of anatomical body parts and an educational program.

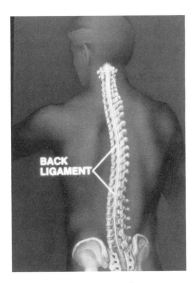

FIGURE Y.   I believe it is important to keep our explanation of body parts in easy to understand terms. Instead of calling this structure by its proper anatomical name (posterior longitudinal ligament), we call it the back ligament.

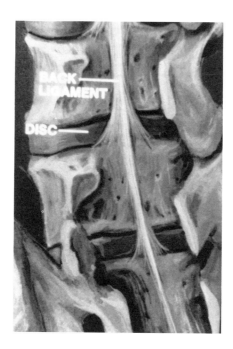

FIGURE Z.   This is a magnified view of the back ligament in the lower portion of the lumbar spine. This view clearly shows the lumbar 4 and 5 discs. It also shows the threadlike structure of the back ligament in this area. Because it is threadlike it is therefore impossible to protect the lower discs when the back is rounded out.

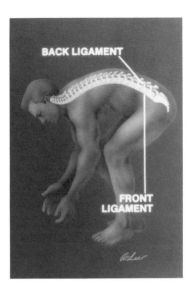

FIGURE AA. Here we show how the back ligament and the back wall of the disc are put on stretch. Most people bend over at the waist to pick up just about everything. They give little regard to what is happening in their back until they hurt. Why should they? They have never been taught any different! Or given rational reasons as to why they should lift differently.

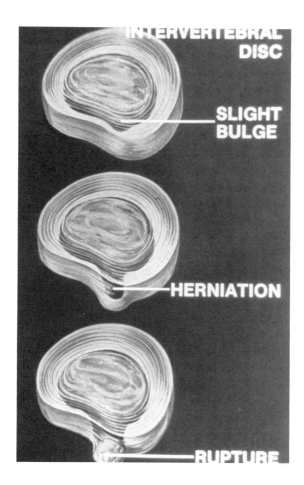

FIGURE BB. This shows the progression of the bulging to ruptured disc. Our explanation is that slipped discs do not exist. Rather, what happens is that the back wall of the disc weakens over time by sitting, lifting, pushing, pulling, and smoking.

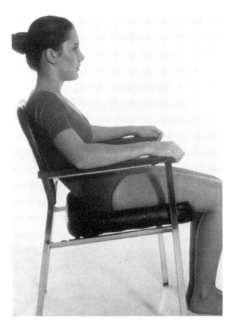

FIGURE CC. This shows typical posture of almost all human beings when you put them in a chair.

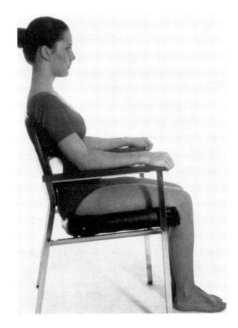

FIGURE DD. A simple way to correct this is by use of a low back support. Our recommendation is to use a low back support 15 to 30 minutes out of every hour you sit. Place this at waist level so that it is in the hollow (lordosis) of your low back.

This is only a sample of what you could use to collect more meaningful data. Make your form specific to your needs. This is a generic from to help you through the thought process.

DATA COLLECTION FORM
LOWER BACK INJURIES

This instrument should be used for the low back only if there is low back pain and reference pain into the hip or leg.

ABC Co.
Lower Back Form

Name of employee

Badge or clock no.

Job title

Time of Injury:	_____ a.m. _____ p.m.
Time into your work shift:	_____ min. _____ hours
(Your pain is in the	_____low back
following body areas)	_____ Hip
	_____ thigh, front part
	_____ thigh, back part
	_____ sitting
	_____ bending over at waist
	_____ lifting bent over
	_____ lifting at waist level
	_____ lifting overhead
	_____ lifting with back locked in
	_____ slipped
	_____ slipped and fell to ground
	_____ lifting bent over and twisted at waist
	_____ pulling
	_____ pushing
	_____ carrying
	_____ standing still
	_____ do not know

What were you doing when this occurred?

Approximate weight lifted: _____ lbs.
How much lifted this shift? _____ lbs.

Season:	_____ summer
	_____ spring
	_____ winter
	_____ fall
Conditions:	_____ sunny
	_____ overcast
	_____ rain
	_____ snow

	hot	_____ above 75°F
	warm	_____ 65-75°F
	cold	_____ below 35°F
		_____ cluttered floor
		_____ grease on floor
		_____ tools on floor

Who was witness to what happened: _____

Have you ever experienced low back pain before? _____ Yes   _____ No

If yes, have you missed time off work? _____ Yes   _____ No.

If yes, do you remember how many days? _____ days   _____ wks.   _____ mo.

Are you on a prescription medication? _____

_____

_____

Have you been through an education low back program? _____ Yes   _____ No

If yes, are you using these new skills daily? _____ Yes   _____ No

Safe job procedure violations, please list:

Could you have used any mechanical devices to assist you? _____ Yes   _____ No

If yes, please describe:

Could you have used the assistance of another employee? _____ Yes   _____ No

Supervisors Summary:

Witness Summary:

Corrective Actions:

1.	accident repeater class	_____ Yes	_____ No
2.	suspension	_____ Yes	_____ No
3.	dismissal	_____ Yes	_____ No
4.	clean job site	_____ Yes	_____ no
5.	help to redesign work site	_____ Yes	_____ No
6.	helper needed	_____ Yes	_____ No
7.	repeat back education	_____ Yes	_____ No
8.	send to M.D.	_____ Yes	_____ No
9.	send to hospital	_____ Yes	_____ No
10.	send to P.T.	_____ Yes	_____ No
11.	nothing	_____ Yes	_____ No

Missed time from work: _____ Yes   _____ No

If yes, fill in: _____ days   _____ wks   _____ shifts

Sent to M.D.   Choice of treatment

_____ medication

_____ analgesics/pain pills

_____ bed rest

_____ other

Sent to P.T.   Choice of treatment

_____ education
_____ exercise stretch
_____ exercise strengthening
_____ ice - rest
_____ instruction in proper
body movements

How many days to return after treatment   _____ days   _____ wks
Employees outcome:

_____ excellent
_____ good
_____ fair
_____ poor

Please describe:

BIBLIOGRAPHY

1. Ahearn, D. Supervisor of Safety, Norfolk and Southern Railroad, Church Street. Roanoke, Virginia.

2. Andersson, G.B.J. *Occupational Low Back Pain.* (New York: Praeger Publisher, 1984) p. 215.

3. Andersson, G.B.J. et al., "A Biomechanical Model of the Lumbosacral Joint During Lifting Activities," *Journal of Biomechanics.* 18: 571-584.

4. Andersson, G.B.J., et al., *Occupational Low Back Pain.* (New York: Praeger, 1984) p. 225.

5. Andersson, G.B.J., et al., *Occupational Low Back Pain.* (New York: Praeger, 1984) p. 218.

6. Apts, Mistral, Porter, Saunders, Selby, and White. "Challenges for the future in injury prevention programs," in *Work Injury Management Prevention.* (Aspen Publishers, 1988) p. 26-28.

7. Apts, D.W. "Back Care Programs: Workers Want to Know, What's in It for Me?," *Industrial Safety & Hygiene News,* p. 15, (November, 1988).

8. Aspden, et al. "Intra-abdominal Pressure and Its Role in Spinal Mechanics," *Clinical Biomechanics,* Vol. 2 (3): 168-174 (August 1987).

9. Ayoub, M. "Control of Manual Lifting Hazaards, PreEmployment Screening," *Journal of Occupational Medicine,* Vol. 24 (10): 751 (October, 1982).

10. Ball, R., Safety Specialist, Marrowbone Development Company, 1989 report to T.R.A.M.

11. Batti'e, et al. "Isometric Lifting Strength as a Predictor of Industrial Back Pain Reports," *Spine,* Vol. 14 (8): 851-856 (1989).

12. Cady, L., et al. "Strength and Fitness and Subsequent Back Injuries in Firefighters," *Journal of Occupational Medicine,* 21 (4): 269-272 (1979).

13. Cain, R.B., Petry, R.L. Investigation of Medical Costs Corresponding to Various Injuries in the Coal Industry and Subsequent Implications of the Need for Ergonomic Research, *Proceeding of the 1984 International Conference on Occupational Ergonomics, Vol. 11,* 465. (To obtain proceedings, write: Human Factors Conference, Inc., P.O. Box 1085, Station B, Rexdale, Ontario, M9V 2B3, Canada.)

14. Cappozzo, et al. "Lumbar Spine Loading During Half Squat Exercises," *Medical Science Sports Exercise.* 17: 613-620.

15. Charniga, A. *1983 Weightliftng Yearbook* Livonia, Michigan: Sporting Press Publishers.

16. Dehlin, O., et al. "Back Symptoms in Nursing Aides in a Geriatric Hospital," *Second J. Rehab. Med.* 8: 47-53 (1976).

17. Delitto, R., Rose, A., Apts, D. Electromyographic Analysis of Two Techniques for Squat Lifting, *JAPTA.* 67 (9): 1329 (September, 1987).

18. Donaldson, et al. "Centralization Phenomenon," *Spine.* 15 (3): 24 (1990).

19. Enoka, R. "The Pull in Olympic Weightlifting," *Med. & Science in Sports.* 11 (2): 131-137 (1979).

20. Farfan, H.F. *Mechanical Disorders of the Low Back* (Philadelphia: Lea & Febiger, 1973, 181.

21. Farhni, H. *Backache Relieved Through New Concepts of Posture,* Springfield, Ilinois: Charles L. Thomas, Publishers, (1966).

22. Farnsworth, R. Annual Reports on LTA's due to back pain, Clinchfield Coal Company.

23. F.E.L.A. *Reader's Digest.*

24. Fisk, J. et al. "Back Schools: Past, Present and Future," *Clinical Orthopaedics and Related Research,* 179 (October, 1983) 18-23.

25. Fitzler, S.I., and Berger, R.A. "1983 Chelsea Back Program: One Year Later. Occupational Health and Safety, (February 1982); pp. 24-26, (July 1982) pp. 52-54.

26. Floyd, D.E., Silver, P. "The Function of the Erector Spine Muscles in Certain Movements and Postures in Man," *J. Physiol.* (129): 184-203 (1955).

27. Garhammer, J. "Energy Flow During Olympic Weightlifting," *Medical Science in Sports & Exercise.* 14, (5): 353- 360 (1982).

28. Garhammer, J. Biomechanical Analysis of Selected Snatch Lifts at the US Senior National Weightlifing Championships, Biomechanics/Kinantroponetry, p. 475.

29. Golding, J. "Electromyography of the Erector Spinae in Low Back Pain," *Postgraduate Medical Journal.* (28): 401 (1952).

30. Gracovetsky, S. et al. "The Mechanism of the Lumbar Spine," *Spine.* 3, (6): 249-262 (1981).

31. Gust, G. "Low Back Pain in Nurses," *Queens Nursing Journal.* (19): 6-8 (1976).

32. Haldeman, S. "Failure of the Pathology Model to Predict Back Pain, Presidential Address," *Spine.* 15 (7): 719-724 (1990).

33. Hall, H. "Back School: An Overview with Specific Reference to the Canadian Back Education Units," *Clinical Orthopaedic and Related Research.* (179): 10-17 (October, 1983).

34. Hart, D.L., "Effect of Lumbar Posture on Lifting" Ph.D. thesis, College of Engineering, West Virginia University (1985).

35. Hult, L. "Cervical, Dorsal and Lumbar Spinal Syndromes," *ACTA Ortho Scand* (Suppl) (17): 1-102.

36. Hult, L. "The Munkfors Investigation," *ACTA Orthop. Scand.* (Suppl) 16 (1954).

37. Irwin, W. Supervisor of Safety, Amax Coal Company. Evansville, Indiana.

38. Kendall, F.P., and McCreary, E.K. *Muscles Testing & Function,* 3rd. ed. (Baltimore: Williams & Wilkins).

39. Kendall, H.O., et al. *Posture and Pain,* (Melbourne, FL: R.E. Krieger Publishing Company Inc., 1952).

40. Kendall, R. "Incentives Encourage Safety Excellence," Occupational Hazards (August, 1984) pp. 39.

41. Klein, A.C., and Sobel, D. *Backache Relief* (Times Books, A Division of Random House Publishers, 1985) p. 150.

42. Lawrence, J.S. "Rheumatism in Coal Miners, Part III, Occupational Factors," *Br. J. Industrial Medicine.* 12: 249- 261 (1955).

43. Jayson, M., Ed. *The Lumbar Spine and Back Pain* New York: NY., 1976 Grune and Stratton, Inc., p. 12.

44. Magora, A. "Investigation of the Relation Between Low Back Pain and Occupation," *Industrial Medicine Surgery.* 41: 5-9 (1972).

45. McGill, S. et al. "Effects of an Anatomically Detailed Erector Spinae Model on L4/L5 Disc Compression and Shear," *Journal of Biomechanics.* 20 (6): 591-600 (1987).

46. McGill, S., Normal, R. "1986 Volvo Award in Biomechanics Partitioning of the L4/L5 Dynamic Moment into Disc, Ligamentous, and Muscular Components During Lifting," *Spine.* 11 (17): 666 (1986).

47. McKenzie, R.A. *The Lumbar Spine: Mechanical Diagnosis and Therapy* Waikanae, New Zealand: Spinal Publications, 1987.

48. McLaughlin, T., et al. "Kinetics of the Parrellel Squat," *The Research Quarterly.* 49 (2): 175-189.

49. Mitchell, R.I., and Carmen, G.M. "Results of a Multicenter Trial Using an Intensive Active Exercise Program for the Treament of Acute Soft Tissue and Back Injuries," *Spine.* 15 (6): 514 (1990).

50. Moffett, et al. "A Controlled Prospective Study to Evaluate the Effectiveness of a Back School in Relief of Chronic Low Back Pain," *Spine.* 11 (2): 120-122 (1986).

51. Nachemson, A.L. "Advances in Low Back Pain," *Clinical Orthopedic.* 200: 266-278 (1980).

52. "Work Practices Guide for Manual Lifting," US Department of Health & Human Services, DHHS (NIOSH) Publications, No. 81-122, March, 1981.

53. Peltier, L. "The Back School of Delpech in Montpelier," *Clinical Orthopaedics and Related Research,* 179: 4-9 (October 1983).

54. Pepper, C. "Quackery: A $10 Billion Scandal," Select Committee on Aging, House of Representatives, 98th. Congress, May, 1984. Comm. Pub. No. 98-435.

55. Pytel, J. and Kamon, E. Dynamic Strength Test as a Predictor for Material & Acceptable Lifting.

56. "Quebec Study," *Spine.* 12 (7 S) (1987).

57. Rissanen, P.M. "The Surgical Anatomy and Pathology of the Supraspinous Ligament and Interspinous Ligament of the Lumbar Spine with Special Reference to Ligament Receptors," *ACTA, Ortho., Scand.* (Suppl) 46, 1960.

58. Roman, R.A., and Shakirzyanov, M.S. *The Snatch, The Clean and Jerk,* (Livonia, MI: Sporting Press, Publishers, 1982).

59. Rose, S., Apts, D., et al. "Functional Assessment Pain Inventory on Underground Coal Miners," in *Coal,* (Chicago: Maclean Publishers, 1982).

60. Rose, S., Shulman, A., and Strube, M. "Functional Assessment of Patients with Low Back Syndrome: An Analysis of Selected Conceptual & Methodological Issues for the Practitioner," Samples can be found in *JAPTA*, p. 672, 1985, *Coal*, February, 1987.

61. Rose, S.J. "Editorial Musing on Diagnosis," *JAPTA*. 68 (11): 1665 (Nov. 1988).

62. Rose, S.J. "Diagnosis and Direct Access," *JAPTA*. 69 (1): 1-2 (Jan. 1989).

63. Sahrmann, S. "Diagnosis by the Physical Therapist, A Prerequesite for Treatment," *JAPTA*. 68 (7): (Nov. 1988).

64. Sander, R., and Meyers, J. "The Relationship of Disability to Compensation Status in Railroad Workers," *Spine*. 11 (2): 141 (1986).

65. Shiro, J.Y. "Differences in Lumbar Erector Spinae Muscle Activity as a Result of Variations in Pelvic Position and Abdominal Muscle Contraction During Bent Knee Lifting of a Load," Masters thesis, Columbia University.

66. Smidt, G. Personal Communcations, University of Iowa, (1986).

67. Snook, et al. "A Study of Three Preventive Approaches to Low Back Injury," *Journal of Occupational Medicine*. 20: 470-481 (1978).

68. Snook, S. "The Design of Manual Handling Tasks," *Ergonomics*. 21 (12): 963-985 (1978).

69. Sorenson, F.B. "Physical Measurements as Risk Indicators for Low Back Trouble Over a One Year Period," *Spine*. 9 (2): 106 (1984).

70. Spillman, M.A. "Effects of a Corporate Health Care Promotion Program," *Journal of Occupational Medicine*. 28 (4): (April 1986).

71. Toner, et al. Back Injury Prevention Training Makes Dollars and Cents, *National Safety News*. (Jan. 1984) pp. 36.

72. Waddell, G., et al. Assessment of Severity of Low Back Disorders and Chronic Low Back Pain, Psychologic Distress and Illness Behavior," *Spine Journal*. 9 (2): 204-213 (1984).

73. Waddell, G. "A New Clinical Model for Treatment of Low Back Pain," *Spine*. 12 (7): 63 (1987); European Edition Suppl. #1, *Spine*. 12 (75): 1987.

74. White A., Panjabi, M. *Clinical Biomechanics of the Spine* (New York: J.B. Lippincott, 1978).

75. White, A., et al. "Synopsis of Symposium on Idiopathic Low Back Pain," *Spine*. 7 (2): (1983).

76. Wiesel, S., et al. "Industrial Low Back Pain," *Spine*. 9 (2): 199-203.

77. Wiesel, S.W., Feffer, H.L., and Rothman, R.H. *Industrial Low Back Pain*, 1985, Charlottesville, Virginia: Michie Company, (1985) p. 6.

78. Wiesel, S.W., Feffer, H.L., and Rothman, R.H. *Neck Pain* (The Michie Company, 1986) p. 15.

79. Isenhagen, S.J. *Work Injury Management and Prevention* (Rockville, MD: Aspen Publishers, 1988,).

80. Zachrisson Forsell, M. "The Back School," *Spine*. 6: 104-106 (1981).

81. Ziglar, Zig *See You At The Top*, Pelican Press.

82. Kapandji, Dr. I. A. *The Physiology of the Joint. 3*, (London: Churchill & Livingston, 1974.) pp. 20-21.

83. Stokes, Ian A.F. "A Critique of Optimum Spine," *Spine*. 12 (5): 511-512 (1987).

84. Rogers, M., Gulf Port Physical Therapy, Mississippi Power Company.

85. Deyo, et al. "Descriptive Epidemiology of Low-back Pain and Its Related Medical Care in the United States," *Spine*. 12 (3): 266 (1987).

86. Mayer, T.G. Journal American Medical Association, 258 (13): 1763-1767 (October 2, 1987).

87. Polatin, P.B., "The Functional Restoration Approach to Chronic Low Back Pain," *Musculoskeletal Journal*. 7(1): 17- 39 (January 1990).

88. McGill, C.M. "Industrial Back Problems-A Control Program," *Journal of Occupational Medicine*. 10(4): 174-178 (1968).

89. Nachemson, A. Clinical Orthopaedics and Related Research. 179: 77-85 (October 1983).

90. Locke, J. *Journal Occupational Health and Safety*. pp. (July 1983) pp. 8-13.

91. Lester, D.M., Assistant Director of Health and Safety, Westmoreland Coal Company — The Accident Repeater, 1982.

92. The Bible, Genesis 1:26.

INDEX

183